LADIES of
HONOR & MERIT

UNIVERSITY *of* PITTSBURGH PRESS

LADIES of HONOR & MERIT

Gender, Useful Knowledge, & Politics in Enlightened Spain

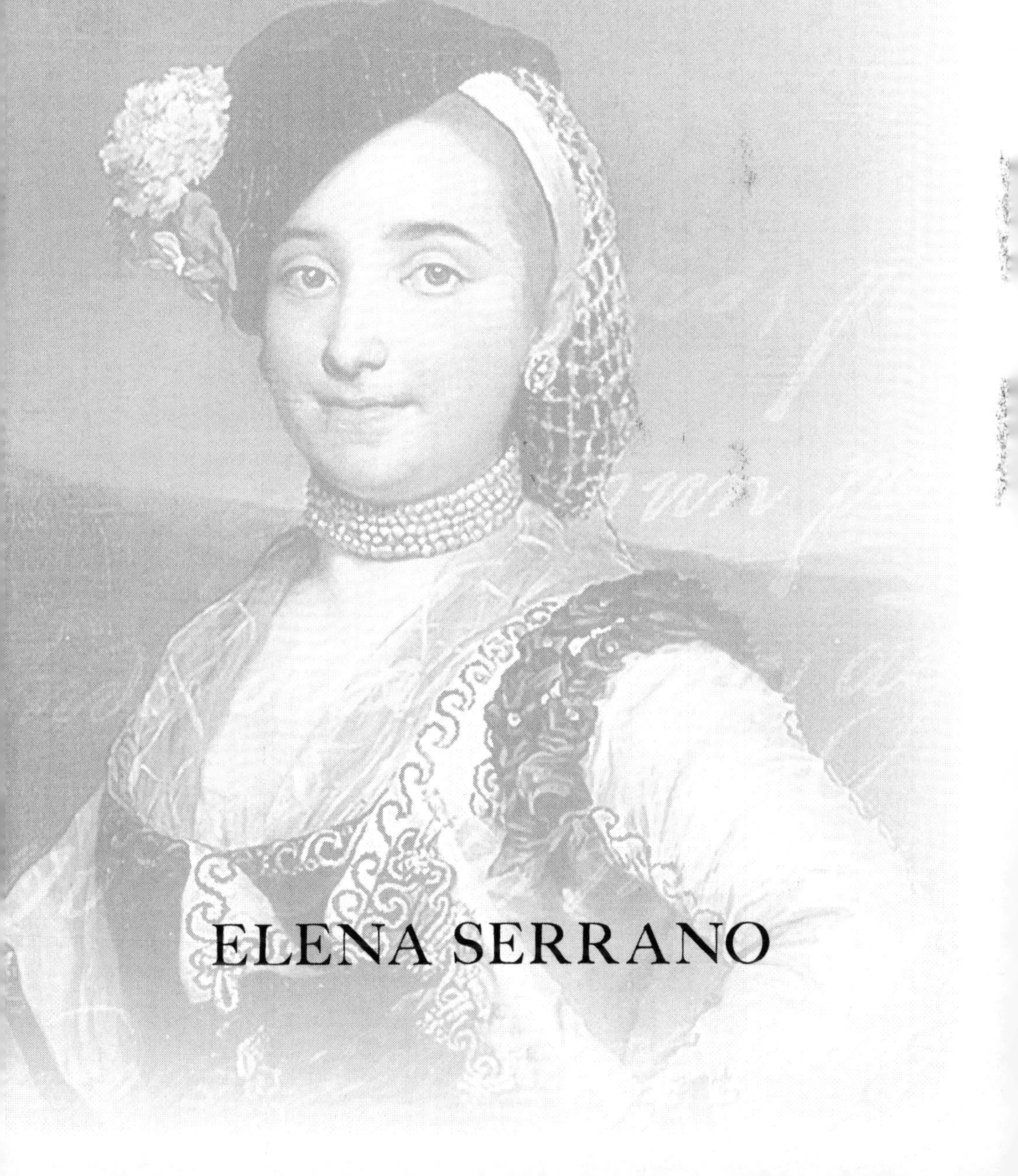

ELENA SERRANO

Published by the University of Pittsburgh Press, Pittsburgh, Pa., 15260

Manufactured in the United States of America
Printed on acid-free paper
10 9 8 7 6 5 4 3 2 1

Cataloging-in-Publication data is available from the Library of Congress

ISBN 13: 978-0-8229-4716-5
ISBN 10: 0-8229-4716-1

Cover art: Portrait of Isabel de Parreño y Arce, Marquise de Llano (c. 1775) by Anton Raphael Mengs, Real Academia de San Fernando; letter written by the Marquise of Llano from Vienna, in which she included a recipe for feeding babies and a sketch of a feeding bottle, Archivo de la Real Sociedad Económica Matritense de Amigos del País; and the insignia of the Real Sociedad Económica Matritense de Amigos del País

Cover design: Alex Wolfe

To my parents, to my daughters

CONTENTS

ACKNOWLEDGMENTS

It is a joy to thank all the institutions and people that have helped me with this project, which has taken more years than it is reasonable. First and foremost, it is my pleasure to thank the Max Planck Institute for the History of Science in Berlin, my second home since I finished my dissertation in 2012. I had the immense fortune of being part of two departments, the one led by Lorraine Daston and the one led by Jürgen Renn. I wholeheartedly thank them for their support throughout all this time. My gratitude to all the colleagues there, who with their conversations, thought-provoking workshops, reading groups, and even casual talks over summer barbecues and movie evenings helped this project take shape. My thanks also to the stupendous staff, in particular Lindy Divarci, who assisted me at every turn. Above all, I want to thank the librarians, always cheerfully helpful even with my most peregrine requests.

Wonderful scholars have accompanied me during this long journey. My deepest thanks to Lissa Roberts, for her unwavering patience in helping me to develop my arguments, time and again, from the very beginning of my career. For her ever wise guidance, I am most grateful to Montserrat Cabré. I am in debt to Antoni Malet for his unflagging support. My heartfelt gratitude to all who took time to read parts of this book and generously offered their advice: Elaine Leong, Christine von Oertzen, Elena Aronova, Anindita Nag, Minackshi Menon, Senthil Babu, Simon Werrett, John Christie, Hansun Hsiung, Elena López, and Mirko Düringer.

I extend my gratitude to Agustí Nieto-Galan and special thanks to all my colleagues at the Centre d'Història de la Ciència in the Universitat Autònoma de Barcelona, in the Instituciò Milà i Fontanals (Barcelona), and in the Institut López Piñero (Valencia), especially to Xavier Roqué, Oliver Hochadel, José Pardo-Tomás, Jon Arrizabalaga, José-Ramón Bertomeu, and Àlvar Martínez-Vidal, who always spared time for kind advice. The Science History Institute in Philadelphia offered me a postdoctoral fellowship, and the Fritz Thyssen Foundation funded a workshop where I discussed part of my research for this book. A chapter was finished while I enjoyed a merry summer fellowship at Sydney University. My thanks to the School of History and Philosophy of Science, where I discussed chapter 1.

I finished the book while participating in the ERC project Circulating Gender in the Global Enlightenment: Ideas, Networks and Agencies (Cirgen). I am grateful to my jolly colleagues there. Particularly I would like to thank Mónica Bolufer, who always offered me her generous and sage advice. In the middle of the pandemic, the silent writing group of York University provided an oasis of calm for revising the book. I would like to thank Michael Thomas Taylor and Irina du Quenoy for their superb editing of the manuscript, and Abby Collier and the whole editorial group of University of Pittsburgh Press that made this project possible. My special thanks to the archivers Fabiola Azanza (Archivo de la Real Sociedad Económica Matritense de Amigos del País), Nacho Díaz-Delgado Peñas, (Biblioteca-Archivo de la Real Academia Nacional de Medicina), and the staff of the Archivo Regional de Madrid for their guidance and support.

Big thanks to my dearest friends, Charo Serra, Ellen Gaske, Murat Türemis, Bernhard Glomm, Pablo Ruiz de Olano, Elena Paulino, and Juan Andrés León who offered me bike rides, lake swimming, Turkish food, consoling beers, and endless nights of conversation, always there for me in the low hours. My thanks to my enthusiastic, dear family: how could I have written a single line without their support? I thank them for their laughter, jokes, and hugs, even from miles away, even when I was too busy to attend to them. This book is for my daughters and my parents.

LADIES of
HONOR & MERIT

Introduction

Women and the Enlightened Campaigns for Improvement

At a time when Rousseauian ideas about women's natural domesticity were in vogue, enlightened politicians and high-ranking women in Spain debated the right of women to join one of the country's main scientific and political institutions: the Real Sociedad Económica Matritense de Amigos del País, or the Royal Madrid Economic Society of Friends of the Country (hereafter Madrid Economic Society).[1] Following passionate debates, women were allowed to set up a female branch of the society. On October 5, 1787, the Duchess of Benavente and nine distinguished gentlewomen solemnly inaugurated the Junta de Damas de Honor y Mérito—the Committee of Ladies of Honor and Merit. From then onward, the women would meet weekly in the Madrid City Hall, keeping to a tight agenda. Within ten years, the Junta had set up a network of over sixty members extending from Tenerife to Asturias and from Austria to Cuba.[2]

This book tells the unknown history of how the Duchess of Benavente, her peers in the Junta, and other high-ranking women actively participated in the new political configuration of the Spanish state, creating, applying, and disseminating "useful knowledge" to improve the conditions of other citizens. Reshaping ideas of feminine erudition and learning as instrumental to the improvement of the country, their vision spread beyond the capital city, fostering other local female societies and inspiring other gentlewomen who sought to have an active role in their social communities.

The Madrid Economic Society was one of nearly five hundred patriotic societies that mushroomed in Europe and the Americas from the 1730s onward, founded on the idea that a better stewardship of natural and social resources could eventually lead to the national improvement—to "public happiness," in the wording of the time. Knowledge, social order, morals, and material culture converged in the activities of the economic societies.[3] Depending on the region, these organizations could be more focused on agronomy, such as happened in the provincial societies of France and Spain; on mineralogy, as it occurred in the German lands and Scandinavia; or, in other places, on the textile industry, fisheries, and timber management.[4] Their members believed that making and disseminating "useful knowledge"—an elastic term that included anything from agriculture experiments to the meticulous observation of nature—would yield better countries and better citizens.[5] They conducted field trials, published journals, commissioned translations of practical manuals, organized prize contests to encourage technical developments, and actively corresponded with other societies. At the same time, within this intermingling of science, morals, and social order, they investigated how to profit from by-products (animal waste for fertilizing land, kitchen ashes and burnt oils for making soap, animal bones for cooking gelatin); decided the fate of communal pastures; set up schools for learning trades and institutions for poor people to be disciplined by labor; and funded orphanages and foundling houses to augment the work force.[6] Reflecting the scope of their aspirations, they called themselves improvers and friends of the country. Typically, they belonged to the ruling classes (the landed gentry, aristocrats, merchants and business men, clergymen, the military, and civil servants) but sometimes also gathered reputed craftsmen and yeomen. However, they included women only occasionally, and we do not know of female branches of economic societies before the Junta de Damas.

This book turns to the outlying case of the Junta de Damas to investigate how women during the Age of Enlightenment negotiated a new political role through rearticulating contemporaneous ideas about femininity in this context of social improvement. Political historians such as Linda Colley have shown how women appropriated Rousseauian ideas about a distinctive feminine nature, and of motherhood as women's primordial social role, as a means to intervene in public spaces. Moreover, they took advantage of

Rousseau's new moral foundations of politics to legitimize their new civic roles, a subversion evident in late eighteenth-century female activism and feminist discourses.[7] Thus, according to Colley, British female activism during the Napoleonic Wars was socially accepted because it was grounded on the "female virtues of charity, nurture, and needlework."[8] In a similar vein, this book argues that in some enlightened Spanish quarters, producing, applying, and circulating useful knowledge in certain areas (mainly rural economy, textile trades, education, and children's welfare) was shaped as the feminine way to contribute to nation-building efforts. It further proposes that the legitimation of women as improvers occurred in a dynamic process of interaction with male reformers, in which ideals of progress and evolving gender identities were mobilized, rearticulated, and negotiated.

In Spain, as in Europe generally at this time, upper-class women played a key role in social imaginaries of progress.[9] Alleviating poverty, increasing population growth, articulating the relationship between households and industries, and regulating the trade of luxurious goods, to quote but a few examples, were issues that benefited from the active involvement of women.[10] Moreover, alongside men, women were responsible for the physical, moral, and emotional welfare of households, which were considered the basis of well-ordered states.[11] Male reformers thus aimed to educate women in the practical knowledge that they believed would make them better stewards of their households, better mothers, and better companions in marriage; at the same time, learned women participated in these educational efforts. This book, then, is a story of negotiations, of rearticulating discourses, of male reformers trying to construct a female collectivity essential to securing their agenda while maintaining their own hegemony, and of women stepping in to the mix on their own terms, with both parties dynamically reframing their respective gender identities.

Yet, while the Junta may be regarded a local phenomenon, the happy conjunction of particular political, social, and biographical circumstances, its members shared a broader Enlightenment spirit of public utility, specifically regarding how upper-class women could be of service to their community, their countries, and even to humanity. The Spanish gentlewomen, the French *femmes* économistes, and the generous patronesses of the Royal Institution in London, to name but a few, all joined the improvement campaign, with pursuits as disparate as inventing new ways of dyeing clothes, gathering statistics about children's diseases, engaging in geology, testing cooking recipes that involved new kinds crops in order to make the latter profitable, and implementing physiocratic ideas in their own landholdings.[12] Thus, in addition to situating Spanish women in the history of knowledge and the Enlightenment, this book contributes to a thriving field that investigates gendered ways of making useful knowledge and constructing expertise and authority.[13]

To be sure, the aim of the Junta was not the intellectual improvement of their members, as it was in the case of the Protestant ladies of the Dutch city of Middelburg—whose Natuurkundig Genootschap der Dames (Women's society for natural knowledge, 1785–1887) met regularly to study natural philosophy—or of the earlier Fair Intellectual Club (1717) in Edinburgh.[14] For the ladies of the Natuurkundig Genootschap, the pursuit of knowledge was a means of getting closer to God and of becoming better mothers and spouses. Around forty wealthy ladies came together biweekly to attend scientific lectures, exchange books on natural philosophy, and carry out experiments. The Fair Intellectual Club seemed to be inspired by contemporaneous male clubs that pursued an intellectual sociability. Their nine young members met regularly in the homes of one of them, presumably to discuss their readings and literary accomplishments.[15]

In contrast, the patriotic goals of the Junta were to be achieved less through their personal edification and more via direct interventions in their communities. Nor were its members public celebrities of erudition (with some exceptions, such as María Isidra Quintina de Guzmán y de la Cerda [1767–1803], whom we will discuss in chapter 1), as in the Italian context. There, in this same time period we find such luminaries as the natural philosophers Laura Bassi (1711–1778) and Cristina Roccati (1732–1797), mathematician Maria Gaetana Agnesi (1718–1799), anatomist Anna Morandi Manzolini (1714–1774), and physician Maria Dalle Donne (1778–1842), among many other celebrated female scientists.[16] Yet, the Junta gathered ladies not only of "honor" (morally upright and belonging to good families) but of "merit," that is, worthy of membership because of their knowledge in the practical issues that the society dealt with.

Indeed, Josefa Amar (1749–1833), Rosario Cepeda (1756–1816), the Countess of Montijo (1754–1808), the Duchess of Benavente (1752–1834), the Marquise of Fuerte-Híjar (1761–1821), and many others women who populate this book were considered ladies of merit.[17] They translated philosophical works, such as Etienne Bonnet de Condillac's *La langue des calculs* and the biography of the political philosopher Count Rumford; they wrote and translated practical manuals (agriculture, education, childcare, medicine); hosted salons, patronized savants, and actively participated in scientific networks.[18] However, their role in the Spanish science have not yet attracted much attention. Apart from some important works on midwifery, nursing, and on the medical writer Josefa Amar, only recently have scholars begun to pay attention their role as circulators and mediators, as the historiographic attention has shifted to other spaces of making science than the traditional ones (universities, laboratories, or male scientific academies).[19]

In the following, I explain the intersections of the three threads of this book, namely, improvement and useful knowledge, gender, and political power.

Improvers

The communality of the "large and motley crew" who enthusiastically sought to be of public utility was insightfully discussed by Lorraine Daston (1999). "Who were the enlightened and how did they get that way?" she asked.[20] As she pointed out, the tricky concept of public utility included, to be sure, securing power and profit for oneself and one's circle. Dealers of natural collections, lecturers of popular scientific courses, the landed gentry of patriotic societies, the enlightened doctors who met weekly in their academies likely sought some kind of material advantages under the banner of utility.[21] Yet, material benefits were often coupled with moral benefits. Increasing agricultural production, for example, was rarely the primary goal of reformists in economic societies; it was always interwoven with the social and material stewardship of local resources and peoples. Scientific education was also meant to form the morals of students; the observation of nature was thought to be deeply edifying.[22] Moreover, the improvement of the unenlightened also presupposed the improvement of the reformer's own self. In Daston's words, to serve the public utility meant "to embark upon a program of improvement, first and foremost, self-improvement. To be educated and public-spirited did not suffice; one also has to learn to think, feel, read, and write in certain ways."[23]

This physiological and spiritual transformation, this reshaping of one's identity to embody a cultural category, was further elaborated by Daston and Otto Sibum in their influential article about the scientific persona, which they define as follows: a "cultural identity that simultaneously shapes the individual in body and mind, and creates a collective with a shared and recognizable physiognomy." They propose an apt image for visualizing how individuals imbued themselves into a persona, that of the mask (*persona* in Latin). Not in the contemporary meaning of *mask*, which hides and suppresses one's "real" identity, but in the ancient Greek and Latin theater meaning, in which wearing a mask made it possible to reach one's full potential.[24] Personae are hence different than stereotypes, social roles, or professions. They are deeply historically situated, being "creatures of historical circumstance," which come and go depending on the times.[25] The scientific personae thus comprise the medieval scholar, the early modern instrument maker, the technocrat, and the naturalist, among many other examples. Personae are collective entities, and it is society that grants significance to this "new ways of being in the world," to these patchy creations of old and new features, sometimes inspired in literary creations but always propitiating new behaviors and meanings in real people.

This book takes this invitation to explore further the relationship between the cultural and the individual identity and suggests considering the case of the eighteenth-century "woman improver." It is unclear whether the woman improver would fully qualify as a scientific persona, but she is certainly intimately related to one of them, the "woman natural philosopher" sketched by Bernard le Bovier de Fontenelle (1657–1757). In Fontenelle's bestseller, *Entretiens sur la pluralité des mondes* (1686), a flirtishly male natural philosopher instructs a marquise about Cartesian astronomy.[26] As Paula Findlen has brilliantly shown, this pair of characters was key in defining the features of both the new male natural philosopher persona (entertaining, witty, and master of "gallant conversation") and the female philosopher—women who could become philosophers in conversation with the men. This literary creation, undoubtedly inspired in real personages, helped to cement further a social reality.[27]

Although the *Entretiens* was not translated to Spanish until late in the century, it was certainly well-known by the elites. One wonders about the extent to which the fictional marquise sparked the taste for astronomy in Spanish enlightened women, such as in the case of Teresa González (ca. 1778), "la pensadora del cielo" (the thinker of the sky); Rita Caveda y Solares (1760–?), who in her *Cartas de una señora a su sobrina* (1800) declared herself fascinated when she raised "a philosophical look to the sky"; or the Countess of Niebla, to whom the engineer Carlos LeMaur (1720–1785) dedicated his astronomical treatise in 1762, because it was the result of "fruitful conversations" with her.[28]

Crucially, the "woman improver" was envisioned to be the opposite of the polite female philosopher. She was set up to defend a different way of researching and a different kind of knowledge worthy to be pursued. While Fontenelle argued in his *Entretiens* that the mission of the philosopher was to unveil the occult mechanisms that account for natural phenomena (famously expressed in his metaphor of the world as a theater in which philosophers were busy behind scenes discovering the pulleys, ropes, and weights that propelled characters and moved scenic elements in seventeenth-century performances), the proponents of the woman improver aimed to demonstrate that the job of philosophers was in fact very different. Namely, they should be less concerned with the "occult mechanisms" than with how to profit from the natural resources noted while attentively observing the theatrical spectacle.[29] This epistemic model was embodied by the Countess of Jonval, one of four characters in the dialogues featured in the hugely successful *Spectacle de la Nature* (1732–1750), authored by Antoine Nöel Pluche and translated into Spanish with great success.[30] The Countess of Jonval could not be more different than the marquise in Fontenelle's *Entretiens*. Rather than looking to the stars, she entertained herself in her gorgeous country house with earthly projects, including collecting shells, drawing detailed colored butterflies modeled in her dry collections, rearing silkworms, caring for her aviary, tending her greenhouse, and of course, enjoying enlightened conversation with her learned husband, a prior, and a young gentleman. The type of knowledge that the countess embodied was construed as a counterpoint to Fontenelle's marquise, arguably one in which women could effectively make significant contributions to society.

Just like the marquise in the *Entretiens*, Pluche's countess was a successful composite of real and fictional elements. She united in her person the cultural practices of fashionable society (collecting, gardening, drawing from nature) and a way of seeking knowledge that successfully negotiated gender boundaries. She might have sparked the imagination of men and women to the point of inspiring emulation, making the idea of a learned women engaged in practical pursuits widely accepted. In fact, although women were not overtly recognized as capable of creating useful knowledge in all quarters, they certainly were in some. While they were not recognized by the reputed members of the French Royal Academy of Sciences, who carefully distinguished their own pursuits of producing "useful knowledge" from those of polite society, in Sweden, women were accepted as knowledge makers in the context of enlightened state-building efforts and the promotion of the utilitarian science.[31] This was also the case among some Spanish reformers, who not only defended the admission of women to the Madrid Economic Society but also brandished the authority of Pluche as proof that this was not a preposterous idea, suggesting that the model of woman improver also worked in other national contexts or social circles.[32] The two models, Fontenelle's and Pluche's, were not mutually exclusive; rather, they offered women socially sanctioned alternative ways to push further the boundaries imposed on their intellectual and political pursuits.

Yet, gender hierarchies were complicated by other categories of difference, in this case, class.[33] While most of the women examined in this book belonged to the highest spheres and had influential salons in Madrid, Spanish reformers mostly belonged to the so-called lesser aristocracy (*hidalguía*). They tended to have paid positions in the bureaucracy or in the judicial system and were often active men of letters who depended critically on patronage for their success. They belonged to economic societies and academies, and frequented other circles of sociability, in particular the salons and *tertulias* run by gentlewomen. Rather than being political radicals, most of them defended the necessity to ally with the monarchy and aristocracy.[34] Thus, as Londa Schiebinger has eloquently argued, although gender might situate women below male reformers within the patriarchal system, women's rank, social influence, and political networks often placed them much higher, complicating the dynamics between the sexes.[35]

Finally, in addition to enlightened state-building efforts, another political circumstance was at play: the recognized utility of female learning for showing off the degree of development of the country. Learned women (in the appropriate measure and form) signaled that the country had overcome a primitive state that mistreated its women, and in the case of Spain, also a Moorish past that supposedly excluded women from society.[36] Let us now briefly examine this context.

The Book in the Historiography of the Enlightenment in Spain

The historiography of the Enlightenment in Spain has been strongly marked by the political, intellectual, and social contexts to which scholars belonged, since the Enlightenment was linked to a specific idea of modernity and its values. Asking whether there had been an enlightened movement in Spain meant asking about the intellectual and economic development of the country, its degree of Europeanization, and its national identity. The "discovery of the Enlightenment in Spain," as Mónica Bolufer Peruga puts it, began in the 1960s, via the work of foreign Hispanists.[37] Jean Sarrailh, Richard Herr, Nigel Glendinning, François López, and others forcefully demonstrated how the state's and Inquisition's censorship mechanisms were never so impermeable as to block the entrance of enlightened ideas; indeed, they showed that new modes of sociability (salons, *tertulias,* coffee shops, academies, societies), of publishing and expressing opinions (the press, the essay, the sentimental novel), aesthetics and sensibilities also entered the country and were refashioned in original ways.[38] In each of the different reigns spanning the Enlightenment (Ferdinand VI, Charles III, and Charles IV), there were periods of greater or lesser openness, depending on international politics and alliances, with intellectual currents sometimes even approaching French revolutionary thinking.[39]

But it was during the 1980s that the research agenda on the Enlightenment intensified, coinciding with the transition to the monarchical democracy headed by the Bourbons following the death of Franco in 1975 (the eighteenth century in Spain was inaugurated by the Bourbons, who ruled the country after the Succession War [1700–1714] unseated the old Habsburg regime). In tune with the changes in the international historiography that understood the Enlightenment as a broader cultural movement not limited to the nucleus of French radical philosophers, scholars discussed the particularities of the Spanish Enlightenment, considered to be more oriented to the "practical" side of the zeitgeist.[40] A further important twist took place when feminist historians explored the role of gender in shaping eighteenth-century culture both in Spain and in Europe generally.[41] They showed, first, the active role of women as agents of change in literature, social customs, and consumer practices, as well as their role in the labor market.[42] Second, they showed how gender hierarchies destabilized supposed universal concepts, such as human nature and human rights.[43] Third, they demonstrated how crucial Enlightenment concepts, such as civilization and sensibility, were defined in relation to gender discourses. In Spain, scholars analyzed for instance the different strategies women used to legitimate themselves as authors and translators, and their importance as *salonnières*, theater spectators, novel consumers, newspaper readers, and educators in the construction of new cultural schemes.[44]

The interdisciplinary research project "Feminist and Enlightenment 1650–1850: A Comparative History" by Barbara Taylor and Sara Knott (1998–2001) crystalized in the terrific *Women, Gender and the Enlightenment*, which with its broad geographical scope and time frame (from seventeenth-century Cartesian feminist to the early years of the nineteenth century) showed not only the deep roots of feminism but also how the differences between the sexes permeated the Enlightenment debates.[45] In this vein, *Eve's Enlightenment: Women's Experiences in Spain and Spanish America, 1726–1839* (2009), with its focus on Hispanic women, analyzed how women in a strong Catholic culture confronted the traditional stereotypes on women's nature and social roles within the contemporaneous social and cultural transformations, in their words, "in the direction of secularism, empiricism and skepticism towards authority," but also full with conflicts and contradictions.[46] In 2020, lastly, the volume on the Hispanic Enlightenment edited by Elizabeth Franklin, Mónica Bolufer Peruga, and Catherine Marie Jaffe highlighted particularly the original re-elaboration of ideas and practices in this context.[47]

To be sure, the rehabilitation of the Spanish Enlightenment occurred in parallel with another historiographic polemic that drove the agenda of historians of science, the so-called polemic of the science. As it well known by Hispanists, the black legend of Spain culminated in Nicolas Masson de Morvilliers's polemical article "Espagne" in the *Encyclopédie méthodique* (Methodological encyclopedia, 1782), in which he infamously asked, "What did Europe owe Spain?"[48] Put simply, some historians of science worked toward demonstrating that, contrary to the international historiographical silence and dismissal, Spain in fact fully participated in European intellectual trends. A different historiographical approach that revealed itself particularly fruitful questioned the center/periphery approach (which focused on the economic circulation of novelties from certain developed countries, mostly in northern Europe, to southern ones). Instead of thinking in terms of processes of passive diffusion, this current understood the circulation of knowledge as a multifarious cultural process and its reception as an active process of appropriation.[49]

Moreover, in recent years, scholars (mostly from the Americas) have proposed that the so-called scientific revolution could not have been possible without the Spanish Empire.[50] José Pardo-Tomás and Juan Pimentel have ironically summarized this latter development in a thought-provoking essay with the telling title "And Yet, We Were Modern: The Paradoxes of Iberian Science after the Grand Narratives." In a nutshell, they write, "Spain reached modernity on the day before modernity was called into question."[51] In addition to a lucid reflection on the role that the history of science played in overcoming the polemic regarding the supposed singularity of Spain, they propose to shift the focus from the Hispanic component and problematize instead modernity and global history. As Lissa Roberts writes, in order to construct a historical map that is extensive in scope and intensive in detail, it is necessary to connect the "specifically local character of individual encounters and the increasingly global networks that both afforded and attributed meaning to these encounters."[52]

This book is situated in these historiographic crossroads. True, it does not tell global histories, one would say that rather the contrary. It looks closely at the material practices of a small group of men and women who considered themselves improvers, mainly in Madrid. In fact, two chapters are confined to the walls of the Madrid Foundling House, while the rest are set in the stuffy office of journal editors, or in the gardens of provincial manors. Yet, the international connections of the members of the Junta and other actors, their intimacy with ideas and materials often received from abroad, demonstrated the global networks they took part in. The wife of a diplomat, the Marquise of Casa Flores, traveled back from Havana and brought with her the roots from the tropical seed *nuevo sagú*, used there to feed the foundlings, with the hopes of alleviating the famines in the Madrid Foundling House. In their travels through Europe, the women examined in this book joined the polite vogue for scientific lectures and were able to learn languages for reading and translating foreign books, actively contributing to the circulation of knowledge. Marina Waldstein, marquise of Santa Cruz and fellow of the Royal Academy of Arts of San Fernando, attended anatomy lessons in Paris, while Catalina de Caso, daughter of a military man with whom she traveled extensively through France, Germany and England, engaged herself in the four-volume translation of Charles Rollin's *Ettudes*.[53]

From the Canary Islands, María de Betancourt, the sister of the famous engineer Agustín de Betancourt, who ended his days employed by the czar of Russia, extracted dyes from the fruits of her orchard in Tenerife, as a way to boost the island's bankrupt silk industry, which the British had hijacked.

María Ruíz de Luzuriaga, the sister of the secretary of the Madrid Royal Academy of Medicine, was urged by the academy to translate a medical treatise on yellow fevers written in Philadelphia, with the expectation of curbing ongoing epidemics in southern Spain, attributed to infected American commercial vessels.[54] Moreover, libraries, correspondence, and other "paper networks" connected the Spanish women who did not travel to the broader world.[55] With a clear focus on the local, this story still hinged on the connection between the local and the global. After all, the optimistic aims of improvers were local and global at the same time. They intended to bridge together the universal—the pursuit of a knowledge useful to humanity—with their motherlands, their regions.[56]

The structure of the book reflects the above considerations. Chapter 1 sets the scene, showing the contradictions that women's education and their social role posed in the Spanish society. Three themes are interwoven here: how learned women were used as proof that Spain had left its Moorish past and entered modernity; how Pluche's *Spectacle* was taken as a model of how high-ranking women could create useful knowledge; and finally, how the debate on the admission of women to the Madrid Economic Society contextualized a debate on the equality of the sexes and on their right to intervene in the public sphere.[57] We will encounter the young Isidra Quintina Guzmán y de la Cerda, the "exceptional woman" who was awarded the first Spanish female doctorate in 1785, and the erudite Josefa Amar y Borbón, who wrote a passionate defense of the intellectual equality of the sexes (1786) to argue in favor of women's admission to the Madrid Economic Society. As already indicated, the debate ended in the creation of the Junta de damas de honor y mérito. Although some women may have wanted to join the male Madrid Economic Society on equal footing, the Junta was still a truly visible platform for developing their patriotic work. Certainly, not all the members engaged with the same enthusiasm and not all attended the weekly meetings; others, however, participated not only in the Junta and its different commissions but also in other societies such as the Ladies of the Jails (Señoras de las cárceles).[58] While some members defended the intellectual and physical equality of both sexes, others were more inclined to grant a different "sensibility" to men and women. Yet, all shared a common sentiment of belonging to a patriotic body of women, which had enormous visibility all around the Spanish world.

Chapters 2 and 3 deal with a harsh topic: the foundling house of Madrid. In 1799 the Junta replaced the former committee of clergymen and civil servants in charge of that institution and assumed total authority over it. The house received three or four babies each day, which at the end of the year amounted to a quarter of the total births in Madrid. Mortality rates among these infants stood at over 95 percent. Along with other hygienic and medical measures, the Junta developed a panoptical paper system for monitoring the health of the foundlings. We will follow the traces that the babies left on paper—in the hefty accounting books, the small parchment slips that were tied to their waist with a code number to prevent their being lost, in the loose sheets of papers for compiling statistics—in order to see how the changes that the Junta introduced in the bureaucratic practices produced medical knowledge. Chapter 3 deals with the thorny issue of the trials in babies for finding a substitute for wet-nurses. The chapter frames the tests within the contemporaneous culture of testing medical treatments on human bodies, researching new diets for the poor, and supposed gender expertise in infant care. It analyzes the power negotiations of the Junta with the doctors of the Royal Academy of Medicine and those who worked in the foundling house, showing how both collectives needed each other in different ways. In chapters 4 and 5, we leave the quotidian urban misery of Madrid and move to the countryside. A key locus in Enlightenment imaginaries of progress was the carefully cultivated farm and—taking the land metaphor further—the careful cultivated person, and in these two chapters we explore the roles of women as improvers of their land and the minds of their children.[59] The book ends in the celebrated gardens of the Duchess of Benavente on the outskirts of Madrid, as an example of how the fictional and the real informed the identity of the woman improver.

Figure 1.1. Portrait of María Isidra Guzmán y de la Cerda. Distributed by the Memorial literario, instructivo y curioso. Copyright © Biblioteca Nacional de España.

1

Enter the Junta de Damas

On June 5, 1785, the young María Isidra Quintina de Guzmán y de la Cerda (1767–1803) was awarded a doctorate in philosophy and belles-lettres by the University of Alcalá de Henares.[1] At just seventeen years of age, she became the first Spanish woman to be awarded a doctoral degree.[2] It was a grand event. A large crowd of more than six thousand people attended her examination and, as was customary, a lavish banquet followed during which orchestral performances alternated with the tolling of the city bells.[3] A silver medal was coined in her name and her portrait hung on the facade of the university for the duration of the three-day exam.[4] Guzmán y de la Cerda won immediate fame: her portrait was published in the prestigious, widely read journal *Memorial literario instructivo y curioso*—a title we might translate as "Bulletin of Instructive and Estimable News"—and her name was added to the lists of "catalogues of exceptional women," that is, women of all periods who had achieved something considered unusual for their sex.[5] Her examination was published in both vernacular and Latin, ensuring a wide readership both in and beyond Spain. Upon her father's request, she was granted admission to the Royal Academy of Language, the Basconga-da Society, and the Madrid Economic Society. Finally, the discourses of gratitude she gave in theses venues were published in the *Memorial literario* as well as in two smaller books.[6]

Isidra Quintina Guzmán y de la Cerda represents the contradictions that the education of high-ranking women—or more generally, the consideration of women's intellect and social duties—posed in the late eighteenth century.[7] Some of the subjects on which she was examined—Latin, Greek, rhetoric, physics, and astronomy—were far from those considered appropriate for women. Moreover, in direct contradiction to female codes of modesty, she displayed her knowledge publicly, in front of large audiences and in male venues.[8] Consider, for instance, her portrait published in the *Memorial literario* (figure 1.1).[9] Sporting a doctoral cap, she is shown busy at her writing desk with a book of philosophy in Latin open on her lap; a quill, an ink well, and a stack of books atop the desk; surrounded by an oval frame crowned with the Latin motto *Causarum Cognitio*, "know the causes." A young woman, who is usually expected to marry and set up a household, is thus represented here as a philosophizing monkish scholar, shockingly crossing customary gender boundaries. Yet, instead of being mocked as a *Latiniparla* (literally: a person—usually a woman—who spoke Latin, but often a derogative term that stood for women who showed off their knowledge while neglecting their domestic and social obligations) or as a *bachillera* (bluestocking), she was honored and received public praise for her outstanding academic achievements.

Furthermore, she was considered both a prodigy—a rarity inspiring admiration—and evidence that Spain had entered modernity.[10] The idea that learned women signaled social improvement had already been expressed in works such as Antoine-Léonard Thomas's *Essai sur les moeurs, l'esprit et le caractère des femmes* (Essay on the morals, spirit, and character of women, [1772]), which was quickly translated into Spanish with great success; but here, on the occasion of Guzmán y de la Cerda's doctorate, the notion was exploited even more emphatically.[11] As the *Memorial literario* put it, "There are no educated nations that could not show a great number of learned and diligent women."[12] Guzmán y de la Cerda's achievement was construed as proof that Spain had overcome its Moorish past and that its women were no longer kept secluded in the home and illiterate. Indeed, France in particular viewed Spain in a predominately negative light during this period. Works such as Voltaire's *Essai sur les moeurs et l'esprit des nations* (Essay on the manners and spirit of nations) and Montesquieu's *Lettres persannes*

(Persian letters) popularly depicted Spain's abject condition, but it was Nicolas Masson de Morvilliers's polemical article "Espagne" in the *Encyclopédie méthodique* (Methodological encyclopedia [1782])—in which he infamously asked "What did Europe owe Spain?"—that excited the most controversy in Spain.[13] Yet, amid European images of Spain as a backward, exotic, and superstitious country, Guzmán y de la Cerda embodied progress and civility. And although in many ways Guzmán y de la Cerda was treated as an exception, she was also heralded as a product of education in a progressive country.

Indeed, Guzmán y de la Cerda was not the only publicly celebrated case of female learning during this period. The young Rosario Cepeda (1756–1816) sat for a three-day public examination in 1768, when she was just twelve years old, in geography, history, grammar, orthography, Euclid's *Elements*, and natural philosophy.[14] The city of Cádiz awarded Cepeda a permanent honorific title, *Regidora honoraria*, which also granted her a salary, and the city's intellectual elites published laudatory poems in her honor.[15] In Zaragoza in 1782, the erudite Josefa Amar y Borbón (1749–1833) was elected as an honorary member of the Aragón Economic Society for her six-volume Italian translation of a work on Spanish literature, which vindicated the place of Spain in European literary history.[16] In Valencia, the aristocrat Pascuala Caro y Sureda (1768–1827) was publicly examined in 1781 and her talent praised in the press. While these women gained fame for themselves and their families, their accomplishments were taken as a sign that their cities had a rich female intellectual life.[17] The case of Pascuala Caro, for instance, was described as "not new for the City of Valencia," suggesting that a good many of her female contemporaries were similarly educated.[18] To be sure, not all high-born women displayed their learning in such spectacular forms, but in all big cities in Spain learned women presided over salons and intellectual *tertulias* (informal social gatherings).[19] The capital, Madrid, had the highest number of aristocrats and thus largest population of *salonnières*; but in peripheral cities, too, such as Valencia, Cádiz, Gijón, Sevilla, La Orotava, or Barcelona, women of letters enriched urban cultural life.[20]

Furthermore, as a product of education, female accomplishments were thought to potentially serve as proof of the equal intellectual capacities of both sexes.[21] The tutors of these women, for instance, used their academic success to promote their pedagogical systems; they presented it as evidence against any claim that only exceptional women could be educated, or that males had superior intellects. For example, Cepeda's tutor Antonio

González de Cañaveras capitalized on her fame in publicizing his method of study as suitable for boys and girls alike.[22]

It is not entirely surprising, then, that the admission of Guzmán y de la Cerda to the Madrid Economic Society forcefully revived an old debate, first begun ten years before when the society was set up, on whether women should be granted membership as a general rule and not as exceptions.[23] The polemics reached beyond the society itself into the public sphere, where discourses in favor of or against women were published in the press and translated into English, Italian, and French.[24] The debates surfaced tensions underpinning society at that time. In particular, they raised questions about the equality of the sexes, both in their intellectual capacities and rights; about what the social and political functions of high-ranking women should be; and on how to co-opt female elites in carrying out a national reform that was seen as urgently necessary, without at the same time giving them too much power. As Amar y Borbón ironically put it, allowing women to have equal footing in the Madrid Economic Society would mean "equating women with men, giving them a seat in their gatherings, and debating serious issues with them—things that seem disruptive and even crazy."[25] The debate ended in 1787, when King Carlos III decided against admitting women to the ranks of the Madrid Economic Society and instead create a female branch of the society, the Junta de damas de honor y mérito (The Committee of Ladies of Honor and Merit), in an arrangement that appears to have been unique in the European economic societies at the time.[26]

Focusing on these debates, this chapter will explore how the role of women in the production of "useful knowledge" was discussed in Spain. Feminist scholars have characterized the Junta as a female charitable organization that enabled women to intervene in civic spaces and exercise a political power of sorts. The historian Mónica Bolufer Peruga has convincingly argued that the economic societies delimited a new political space situated between the Iberian medieval representative assemblies (*Estamentos*), which were inflexibly defined by their masculinity, and the liberal courts that were still to come. Because the economic societies were not defined by a restrictive masculine identity, the women of the Junta were able to develop a feminine political identity—what Theresa Ann Smith has called "the female citizen."[27]

This chapter adds to these conversations, framing the Junta within the full context of the improvement campaign, in which scientific knowledge had a crucial role. While it is true that economic societies shared some features with charitable societies—their service to the community, patriotic and humanitarian character, and will to reform society—equating the Junta to a philanthropic society nevertheless overlooks both the fundamental role that knowledge played in the economic societies' reformist activities and how this knowledge was practiced and understood by eighteenth-century actors.[28] Like the approximately five hundred economic societies that mushroomed in Europe at the end of the eighteenth century, the Madrid Economic Society was founded on the belief that laypeople (as opposed to savants or experts) could also contribute to local prosperity through technical innovation and the circulation of what was called "useful knowledge." The members of these societies called themselves improvers, patriots, and "friends of the country." Typically, they brought together the local learned elites—clergymen, gentlemen, landowners, and members of the military and liberal professions.[29] This chapter aims to show how well-off women were considered by a substantial part of this crowd of improvers to be legitimate actors in the production and circulation of useful knowledge despite the limitations imposed by their gender. It further contends that the Junta must be understood historically as functioning within this improvement movement. The issue is no minor question of terminology: including the knowledge behind the social reform that the Junta undertook within the framework of analysis allows us to uncover gendered ways of making useful knowledge.

Along with the idea that the Madrid Economic Society delimited a new political space still ambiguous in its masculine identity, it is important to highlight that, unlike other European countries, Spain did not have a central academy of sciences populated by prestigious natural philosophers. The Madrid Economic Society performed some of the functions of these academies, serving for instance as a technical consulting body to the government, but was by no means as powerful as France's Académie des Sciences or Britain's Royal Society of London. Nor did it share their elitist character. The idea of admitting women thus did not frontally conflict with the conception of the expert as a man, or of scientific pursuit a manly enterprise.[30] On the contrary, those members of the Madrid Economic Society who defended the admission of women considered the latter to be active agents of change; they believed in women's capacity to alter social behavior, consumer habits, and attitudes toward work and to create and circulate useful knowledge—things considered of the utmost importance for improving the country. As we will see in the following discussion, their ideas were not only based on theories of political economy theories but were also shaped by contemporary female literary models, which convincingly featured gentlewomen engaged in making useful knowledge.

Politics, Gender, and Useful Knowledge

It is well known that male economic and scientific societies sometimes admitted "exceptional" women to their ranks. Scholars have interpreted their admission as a means by which male societies could highlight their masculinity.[31] Yet, as Lisbet Koerner has argued, philosophers assigned an important role to women in empowering nation-states.[32] Women were occasionally chosen as honorary members of scientific societies because their achievements were considered important for stimulating national growth and national identity. For instance, in 1748 the Kungliga Vetenskapsakademien (Royal Swedish Academy of Sciences) elected Countess Ekeblad for her experimental work on extracting starch from potatoes, which led to their use in making aquavit, flour, and wig powder. A similar rationale underpinned Josefa Amar's election to the Aragón Economic Society in 1782, as her translation of the Jesuit Francisco Javier Llampillas's work was thought important for defending the country from foreign attacks against Spanish culture.[33] Similarly, the Marquise of Cerralbo was appointed a member of the Ciudad Rodrigo Economic Society for her generous funding of the local hospital.[34]

But even beyond these cases, political economists believed that alliances with influential women were crucial to facilitating change. Women were understood to have the ability to deeply sway society and thus to change patterns of consumption, social behavior, and moral values, albeit in contradictory terms. On the one hand, women embodied the power to civilize: good taste, polite conversation, and upstanding morals were supposedly spurred on by women's presence and conversation. On the other, they were understood to be inconsistent, whimsical, and driven by fashions. A good political economist, it was said, should put both sides of female nature to work for the communal good.[35] For instance, this argument suggested, if trendy ladies consumed national products instead of imported luxurious goods, others would follow suit, eventually balancing the national debt.[36]

The aim of economic societies to contribute to and foster public happiness depended crucially on the gendered roles of men and women. Issues such as how to manage the poor, how to increase the working population, how to articulate the relationship between households and industries, and how to regulate the trade of luxurious goods, to mention only a few, were understood in terms of gender.

This can be seen, for instance, in the approach of economic societies to administering poor relief. Bound up with household roles, class, and religion, charity was traditionally central to the identity of upper-class women.[37] Thus, the changes that reformers pursued, such as setting up working houses, regulating alms, and classifying which poor deserved aid and which did not, required women to agree with these new ways of making charity effective. Women were also fundamental to plans for increasing textile production. The political economist Pedro Rodríguez de Campomanes (1723–1802), a stateman and major player behind the boom in economic societies in Spain during the 1770s, favored outsourcing some parts of textile production to households, which meant transferring both the work and the responsibility for this activity to women.[38] Moreover, the belief that virtuous households laid the ground for happy states—in other words, that private virtue was reflected in the public order—put contemporary thinkers in a thorny dilemma: how could society be reformed while respecting individual freedom? One of the most influential political economists in Spain, the German Johann Heinrich Gottlob von Justi, admitted that controlling households was not only "impossible in itself" but also impossible "in terms of good principles."[39] Similarly, Campomanes argued it was unethical for officials "to meddle in the secrets of families."[40] High-ranking women, however, offered a way out of this dilemma, as they would be able to monitor and direct low-class households through their own charitable activities and supposed moral influence.

Given the central role political economists accorded to women as agents of change, it is not surprising that in 1775 (ten years before the exceptional admission of Guzmán y de la Cerda), one of the first issues the Madrid Economic Society debated upon its founding was whether to admit women to its ranks. Campomanes supported the measure, largely because of the benefits he believed would accrue to textile industries through women's influence. He thought female members could, for example, stimulate well-off women to undertake textile production at home, thereby involving their daughters and their servants in the endeavor, organizing home workshops, and, above all, overseeing poor women's households. Another interesting argument raised at the time stemmed from the reported beneficial influence of well-to-do French women on encouraging the study of natural philosophy among men. One member of the Madrid society, José Manuel Marín y Bordá, pointed out how their zest for "curious books" had spurred the production of new knowledge as well as increased its circulation: "When the ladies of France made the reading and studying of curious books fashionable, they inspired the same taste in men who adored them; because the example [set by] these noted [ladies] and the wish to please them spurned the men to enrich themselves with news and essays to content them during their visits."[41] Marín believed that if women were admitted to the Madrid Economic Society, their example would instill the same appetite for studying in Spanish men, which eventually would lead to national prosperity. Moreover, women's engagement in issues discussed at the society would generate a positive attitude toward agriculture and the arts in the men around them. Similarly, it would encourage other women to engage in rural economy when they saw their discoveries—"an improvement, method, practice, or any other useful thing"—widely praised by economic societies.[42]

As we will see in the next section, in seeking to demonstrate that women could make useful discoveries, Marín drew on the authority of the Abbé Noël-Antoine Pluche, whose female character in the *Spectacle de la Nature* (1732–1750), the Countess of Jonval, had a prominent role in this encyclopedia-like compendium, which was translated into Spanish during this period with outstanding success.[43]

Women Who Converse about Science

The afternoon conversations of the Countess of Jonval with her husband, the parish priest, and a young gentleman from Breuil were famous all over Europe. While strolling through her garden on the outskirts of Paris, the countess amuses her companions by expounding on the natural life of silkworms and bees, the skillful ways in which siskins nested in her aviary, and the deplorable behavior of female ostriches toward their offspring.

These lively conversations form the narrative of the Abbé Pluche's *Le Spectacle de la Nature, ou Entretiens sur les particularités de la Histoire naturelle qui ont paru les plus propres á rendre les jeunes-gens curieux* (Spectacle de la nature: Or nature display'd: Being discourses on such particulars of natural history as were thought most proper to excite the curiosity of youth).[44] Almost forgotten today, the *Spectacle* was one of the great European best-sellers of the eighteenth century. Written in the tradition of natural theology, the *Spectacle* sought to induce piety and religious fervor in its readers through the observation of the wonders of nature. Ann M. Blair has argued that its historical significance has been overlooked largely because its religious character and emphasis on the limits of human understanding do not sit well with a certain image of the Enlightenment as a revolutionary intellectual enterprise.[45] Yet the *Spectacle* was listed in five hundred private library catalogues printed between 1750 and 1780 and was the fourth best-selling book between 1750 and 1780, surpassed only by Bayle's *Dictionary*, Marot's *Oeuvres*, and Buffon's *Histoire Naturelle*.[46] It was translated into all the main European languages and repackaged in abridged formats, inexpensive editions, and luxurious ones, such as the Spanish quarto edition.

In eight volumes, the *Spectacle* explains all that Pluche, a former tutor in a gentle family, considered important for the education of a young gentleman: from the natural history of insects to agriculture, physics, crafts, and theology.[47] Despite its breadth, the *Spectacle* proved successful in reaching a wide audience.[48] One of the reasons might be that the topics were arranged according to increasing complexity, from natural history to physics to theology—or, as the saying went, from insects to God—thus appealing to readers across a wide range of interests. Pluche also provided a detailed bibliography, so eager readers could always expand on the different topics.[49] Undoubtedly, its fantastic engravings also played their part. These included not only classical etchings from well-known natural history manuals but also sketches from garden catalogues, fashionable Dutch floral paintings, and engravings of the arts and trades.[50] But probably more importantly were the characters, "worldly people" as Pluche puts it, to whom readers could feel close.[51]

However, in obscuring the *Spectacle*'s significance historically, the importance of the Countess of Jonval has been lost as well. The evidence suggests that she played an important historical role as a model, a scientific persona, for what we could call the woman improver.[52] Scholars have instead focused primarily on the female character featured in the earlier best seller, *Entretiens sur la pluralité des mondes* (1686, published in English in 1687 as *A Discourse of the Plurality of the Worlds*).[53] Written by Bernard le Bovier de Fontenelle (1657–1757), the permanent secretary of the Académie des Sciences, the *Entretiens* narrates the nocturnal conversations on Cartesian astronomy between the Marquise G. and a gallant natural philosopher. The *Entretiens* triggered an onslaught of instructive dialogues between a male tutor and a female pupil, including works such as John Harris's *Astronomical Dialogues between a Gentleman and a Lady* (1719), Benjamin Martin's *The Young Gentleman and Lady's Philosophy* (1759), Francesco Algarotti's *Il Newtonianismo per le dame* (1737), Giovanni de Cataneo's *Philosophizing for Beauties* (1753), and Giuseppe Compagnoni's *La Chimica per le donne* (1792), all of which were translated into several different languages and circulated widely.[54] Although these works were inspired by Fontenelle's *Entretiens*, their authors adapted them to their local readers and interests.[55]

Pluche, for his part, suggested that the *Spectacle* was his response to the intellectual arrogance of the philosopher in the *Entretiens*. It is also likely that he intended the Countess of Jonval to represent the antithesis of Marquise G. A married woman instructed not by a flirtatious philosopher but by her husband and through her own self-guided study, the Countess of Jonval was more interested in down-to-earth knowledge than in the heavens. While the Marquise G. learned about astronomy, the Countess of Jonval studied natural history, agriculture, and rural economy.

Instructive dialogues featuring a female character served different, even contradictory, functions.[56] On the one hand, a female interlocutor indicated that the genre was welcoming, easy to understand, and fit for lay practitioners. These instructive dialogues were instrumental in spreading not only knowledge but also moral values and scientific behaviors: how to observe, collect, and hold scientific conversations; use instruments; and draw specimens. On the other hand, such books certainly presented women interested in scientific pursuits. Whether conversing with natural philosophers, buying instruments, ordering collections of natural objects, or observing nature, the deeds of literary and real women shaped each other: authors drew their personages from real women and literary figures in turn stimulated flesh-and-blood female readers, who then also inspired authors and publishers.[57] As Paula Findlen put it, "a bookish idea helped to solidify a social reality."[58]

Yet, instructive dialogues also served to delineate boundaries between experts and nonexperts, to hierarchize certain types of knowledge above others. Fontenelle's Marquise G. is curious and quick but lacks the mathematic skills and constancy for carrying out careful observations over long periods of time. In placing a conversation about astronomy in a romantic setting and making sure that the Marquise G. lacks any firm basis in mathematics, Fontenelle skillfully splits and genders two spheres of knowledge: the masculine, useful, mathematic knowledge of the Académie des Sciences and the feminine and polite knowledge of the salons. Fontenelle is clear as to who was entitled to make science: only (male) academics.[59]

Pluche, in contrast, blurs the boundaries between who is entitled to make knowledge, rearticulating which knowledge is useful to pursue. He shifts away from the traditional gender hierarchies in which women were the pupils and men the tutors, balancing the conversations between the four characters. The Countess of Jonval contributes on an equal footing with her companions to the teaching, although each character had their own particularity, one might even say embodied a different approach to the knowledge of nature. For instance, the count represents the learned gentleman: he has a cabinet of curiosities and collects and classifies minerals. He displays a deep knowledge of the subject. The prior represents the way men of religion should approach natural philosophy: the fellow of a local scientific society, he always stresses the moral aspect of an observation. On the contrary, the chevalier represents the curious young lad of quick imagination. Finally, for her part, the countess embodies the ideal learned gentlewoman, whose knowledge helps her to govern her household with prudence.[60] She appears on stage at the precise moment the three men are about to talk about silkworms, in which she is an expert (as silkworm breeding was seen as an appropriate duty of women). She lectures on the natural history of moths, whose behavior she observed in order to preserve the family furniture, and on other useful skills, such as how to keep an aviary, cultivate flowers, and grow strawberries in a greenhouse during the winter. Yet, of particular interest in relation to the movement for improvement, the *Spectacle* argued for a different kind of knowledge, and a different kind of practitioner than Fontenelle in his *Entretiens*: one who is less interested in "hidden causes" and more in "useful knowledge."

Fontenelle popularized the classical metaphor of the world as a "theater of machines," in which the role of the natural philosopher was to unveil the hidden machinery that moves the scenery.[61] In contrast, for Pluche philosophers should concentrate on "what is possible to know." He argued against Cartesians, Newtonians, and other "mathematicians," as he called them, who used tourbillons, springs, and gravitational forces to explain the unknown.[62] Pluche's position is plainly stated in a long letter sent by the prior to the chevalier at the end of volume 1. Now back in his homeland and planning to continue his study of nature, the chevalier is unsure of how to go about it and asks the prior for advice. The latter's answer is that one should recognize his own limits; trying to understand everything would yield the same perplexity and doubts as not trying at all: "With respect to all created Things that rise to our View, there are but three Particulars wherein we can fix a Determination: One must be a Resolution to know nothing; the second, a Desire to comprehend the Whole; and a third, an Inclination to search after and improve to the best Advantage whatever we are capable of knowing. The first Determination is a piece of Indolence that runs into mere Stupidity; the second is a Temerity which is constant; and the third is a Resolution of Prudence."[63]

Pluche does not distinguish between the capacities of women and men to yield profitable knowledge. Together the four characters form a kind of "Académie des sciences" in which the countess is at the same rank as her male peers. The prior explicitly names the countess a "true Physicien" because of her capacity to observe nature: "Madam, the merit of physicists [*Physiciens*], among whom we now count you, does not always consist in guessing difficult things; but in opening one's eyes to what others do not see, and which they most often lie before their feet. Nothing is rarer than people who think and reflect."[64] In sum, the way Pluche understood the kind of knowledge that is worth pursuing opened the door to women just as Fontenelle has done before, but this time wider and in more equalitarian terms.

The Spanish Translation of the *Spectacle*

The significance of useful knowledge and the role of women in producing it was even stressed in the Spanish translation of the *Spectacle* by the Jesuit Esteban de Terreros y Pando. First, Terreros tweaked the title by adding to it the adjective "useful": *Espectáculo de la Naturaleza, ò Conversaciones a cerca de las particularidades de la historia natural, que han parecido mas a propósito para excitar una curiosidad útil, y formarles la razón à los jóvenes lectores*. Second, he deviated significantly from other European translations by adding some fifteen hundred footnotes, most of them about local techniques. As I explained elsewhere, Terreros envisioned the *Espectáculo* as an encyclopedic reference work for the arts and sciences, which the nation lacked and which dovetailed with the Spanish monarchy's reformist agenda.[65] To help his readers understand and apply the book's lessons, he sent hundreds of queries to craftsmen and naturalists to check the accuracy of his translation of French terms. In addition to his translation of the *Spectacle*, Terreros compiled what was in essence the first dictionary of the arts and sciences in Spanish—a fact suggesting that he understood his translation, too, as a tool for modernizing the country.[66]

It was thus in keeping with Terreros's aims that he dedicated his *Espectáculo* to the queen and declared that the goals of both the book and the monarchy—public happiness—were the same.[67] He stressed his concern for women's education and the need to cultivate their literary taste.[68] He went so far as to detail the subjects everyone—independent of their sex—should receive instruction in, such as logic, history, experimental physics, and religion.[69]

His painstaking efforts to translate accurately every fragment of information in Pluche's work were praised by Benito J. Feijoo (1676–1764), one of the most influential Spanish philosophers and ardent advocate of the intellectual equality of women, who considered Terreros's translation a hallmark of Spanish literature.[70] Moreover, the literary critic Juan Sempere y Guarino claimed the *Spectacle* was responsible for sparking women's interest in natural history and the crafts: "Even upon the *estrados* [a classical elevated platform with cushions in the room where women met] and among ladies, it became fashionable to talk about the natural history of animals, plants and minerals, and about crafts and manufacturing."[71] Sempere y Guarinos's claims regarding the work's influence among Spanish women must be read with caution, however. He was writing to offset foreign criticism of Spain for its backwardness and described the learned female population in a light intended to demonstrate the country's modernization. Nevertheless, the fact that a female character played such a prominent role in the *Spectacle* may well have inspired real women.

It is no wonder, then, that Marín presented Pluche's work as decisive evidence in favor of supporting women's admittance to the Madrid Economic Society in the initial debates in October 1776. These were shut down indefinitely, however (and we still do not know why), resuming only ten years later with considerable vigor following the admission of Isidra Quintina Guzmán y de la Cerda to the society.[72]

The Debate of 1786

Shortly after the Madrid Economic Society admitted the young Guzmán y de la Cerda, it elected its second female member, the formidable Duchess of Benavente, María Josefa Pimentel Téllez-Girón (1752–1826), who was also the wife of the then-president of the society. Their admissions reopened the debate on the women question, and this time the deliberations spilled over into the public sphere, as discourses for and against were published in the *Memorial literario*.[73] In response, Josefa Amar y Borbón sent her *Discurso en defensa del talento de las mujeres, y de su aptitud para el gobierno, y otros cargos en que se emplean los hombre* (Discourse on the defense of women's talent, and their aptitude for government and other positions in which men are employed) to the Madrid Economic Society, which was read on June 24 and also published in the *Memorial literario*.[74] Another woman, Madame Levacher de Valincourt, wrote a long article that appeared in the journal *Espíritu de los mejores diarios que se publican en Europa* (A digest of the best newspapers published in Europe).[75]

As feminist scholars have shown, male advocates of women did not always agree on what women's participation should consist of. Their differences ranged from small variations on minor issues (whether women should pay a membership fee as men did), to disagreements on more fundamental points (whether women should attend men's meetings or rather meet in a private venue, or what the criteria should be for selecting female members). All, however, raised the issue of what roles women should play in society. For instance, the writer and statesman Melchor Gaspar de Jovellanos believed women could assist male members on certain issues in which women had more expertise (textile manufacturing or girls' education), but that "their prudence" would keep them away from the meetings of the Madrid Economic Society.[76] Juan Antonio Hernández de Larrea, a member of the Aragón Economic Society, defended the admission of women, stressing how they could contribute to rural economy in the way that Pluche had imagined: "Not digging, not ploughing, not carrying heavy loads as our women in the mountains do, but [the ladies] can and must direct and take care of harvesting and spinning silk as the Chinese empresses do; they can cultivate flowers and vegetables; they can rear birds, prepare seeds with bleach solutions to make them fertile, and there are thousands of other rustic labors that women can do to foster agriculture."[77] In contrast, Francisco Cabarrús (director of the National Bank of San Carlos and member of the Consejo de Hacienda) categorically opposed women's admission and viewed elite women as having no obligation other than being mothers of happy extended families.

We do not know what the women who aspired to join the Madrid Economic Society thought, except for Josefa Amar y Borbón, who very clearly expressed her thoughts about the intellectual equality of both sexes and the right of women to join such societies.

The Voice of a Learned Woman

In the summer of 1786, Amar y Borbón entered the debate in reaction to Cabarrús's forceful objections against the admission of women.[78] Her contribution offers a rare and important source for understanding what might have been the perspective of learned women on this issue. Married to a judge on the royal court in Saragossa, the daughter and niece of court physicians, Amar y Borbón belonged to the gentry on her maternal side and to the emerging liberal professional class on her paternal side and by marriage. Already an honorary member of the Aragón Economic Society, she felt it was her moral duty to make a case for the general admittance of women to economic societies: "Women should vindicate their cause, because on this occasion their silence would confirm the opinion . . . that they do not care about or attach importance to serious business."[79] Amar y Borbón's *Discurso en defensa del talento de las mujeres* (Discourse in defense of women's talent) is one of the most interesting and touching eighteenth-century texts on the rights of women to participate in public affairs. Passionate and learned, she delineated her defense of female membership with a list of thirty-four points, dismantling one by one the arguments advanced by the society's male members.

She began by spotlighting the paradoxical and censorious behavior of men toward women. Men whined about women's lack of instruction, but they neglected to provide it. Men flattered women while at the same time blaming them for social mishaps. Men belittled women in public yet still sought their company. She adroitly argued that the fact that men had such strong opinions on the influence—bad or good—of women demonstrated just how powerful women were. In points five and six, she complained about the universally unfair social condition of women, either as slaves in Muslim countries or dependent on men in Western ones, which discouraged any attempt at intellectual effort: "Women know that they cannot apply to any position or public reward—that their ideas will have no longer journey than the walls of a house or a convent."[80] In points seven through nine, she turned the accepted religious arguments against women's greater participation in society on their head, arguing that far from signifying her dependence on Adam, God's creation of Eve signaled Adam's need for companionship and society. Eve's curiosity was proof of her talents because what she longed for was knowledge. The fact that women were rendered subservient to men after their expulsion from paradise paralleled cases in which ingenious men were forced to obey others who were less educated but more powerful.[81]

Points eleven to twelve are perhaps the most revealing in terms of the sort of woman Amar y Borbón believed was most suitable to be a member of economic societies. In order to gather proofs of women's capacities, she turned to the long tradition of using catalogues of illustrious women (a centuries-old genre that recorded all types of women, including saints, martyrs, goddesses, soldiers, erudite women, and so forth that were used in the disputes over women's suitability for various societal roles). Amar y Borbón took great care in selecting which women to take as her exemplars, singling out particularly learned women.[82] In addition to noting several contemporary French women well-known for their pedagogical books, such as the writer Mme de Genlis, she referenced the empress of Russia, Catherine the Great, widely known for presiding over the Imperial Academy of Sciences and Arts in St. Petersburg and supporting the ideals of the Enlightenment.[83] Amar y Borbón stressed the fact that if Catherine was capable of presiding over an academy of sciences, women could be members of a "body" that was "not devoted to such abstract matters."[84] Women of merely average intelligence could easily serve the patriotic mission of the Madrid Economic Society: "To compare foreign works with ours to see what can be improved, to make new inventions in the arts and improve the known ones . . . although these are things that demand meditation and staying up to date, they are not such abstract matters that a woman of normal talent cannot understand [them]."[85]

Yet in spite of the eloquence of all those who had advocated for women's admission to the Madrid Economic Society, it ultimately did not grant them membership. Instead, as we know, a separate committee, the Junta de damas de honor y mérito, was established. Their mission would differ from the male economic society's, narrowed to what was considered the feminine sphere. As King Carlos III ordered, the Junta would "promote education, good behavior and respect for work, and the rejection of luxury products."[86] Notwithstanding Amar y Borbón's ardent defense of the equality of the sexes, men and women were assigned different roles according with the prevailing gender model.

The First Endeavors of the Junta

The first meeting of the Junta was held on October 5, 1787, in the Madrid Council Hall.[87] From then onward the women would meet there every Friday afternoon—not to coincide with the Madrid Economic Society's Saturday meetings.

During the three-hour-long first meeting, the Junta elected officers for a variety of posts. As president, they chose the Duchess of Benavente, and as secretary the Countess of Montijo. But probably the shrewdest move of the Junta at this point was to propose that princesses be made members, specifically naming Princess María Luisa, who was married to the future Carlos IV, and the *infantas* (the title given to the daughters of the king).[88] This meant that from henceforth the members of the Junta could turn to the queen when they needed financial or other types of support for their enterprises.[89]

The first president, the Duchess of Benavente, is better known today by the title she acquired through her marriage, the Ninth Duchess of Osuna.[90] She was one of the most influential Spanish women of her time, having received the aristocratic title of *grande de España* four times. Following her marriage to the Duke of Osuna in 1771, she owned most of the land in the south of Spain as well as in Salamanca, Valladolid, and Valencia.[91] Her family had amassed one of the largest libraries in Spain, with more than thirty-five thousand volumes, and her salons at the Palacio de la Vega in Madrid and at the manor El Capricho on the outskirts of the city hosted artists, writers, and savants.[92] Guests enjoyed the latest compositions of Joseph Haydn, as well as new plays by Ramón de la Cruz (1731–1794), Leandro Fernández de Moratín (1760–1828), and Tomás de Iriarte (1750–1791), in which the duchess sometimes performed. The spirit of this "femme rare et sublime," as she was called at her death, was probably best captured by Goya in his portrait of 1785 (figure 1.2).[93] Dressed in the latest Marie Antoinette fashion, she is depicted with a confident gaze directed at the viewer. She leans against a parasol in the same way that soldiers rest against their swords

Figure 1.2. Portrait of María Josefa de la Soledad Alonso Pimentel (1752–1834), condesa-duquesa de Benavente, today also known as the Duchess of Osuna. Goya (1785). Fundación Bartolomé March, Palma de Mallorca. Public domain.

and politicians against their batons, exhibiting not only her charm but also her power.[94]

The first secretary, the Countess of Montijo, María Francisca de Sales Portocarreño (1754–1808), had strong ties with reformers such as the already mentioned Gaspar Melchor de Jovellanos, the poet Meléndez Valdés, the botanist and editor of a journal for agriculture and arts Juan Antonio Melón (chapter 5), and Ignacio Ruíz de Luzuriaga, the secretary of the Royal Academy of Medicine. The duchess herself translated a French treatise on marriage in her youth.[95] A very active member in the Junta (she would be one of those directly responsible for the reforms in the Madrid Foundling House), she would also set up another female society, the Señoras de las Cárceles (The Ladies of the Jails) with the aim of caring for the female prisoners in the three jails of Madrid.[96]

The future members of the Junta belonged mostly to the aristocracy and were often spouses and daughters of ambassadors, diplomats, and other high civil servants. They had thus extended circles of sociability and influence, either through their duties as spouses of public figures, their kinship networks, or through patronage.[97] Their networks could reach far and wide, as they would accompany their husbands in their travels abroad.[98] In addition, they were learned women, as the statutes of the Junta required. Article III of the society's founding charter stated that all members must "possess a good education, exhibit proper conduct, and have strong instruction in the subjects of the committee."[99] New members would be elected by secret ballot to ensure that only those who met the standards would become members. The statutes also stipulated that all members could put forward any "useful proposal," and some of them would "carry out all the experiments and communicate all the findings that [they] were commissioned with."[100] The best-known members of the Junta included the Marquise of Espeja, born Josefa de Alvarado Pacheco (b. 1803), who in 1785 translated a treatise on moral philosophy by the Italian Zenotti, as well as Condillac's posthumous *La langue des calculs* in 1802.[101] They also included the Marquise of Fuerte-Híjar, born María Lorenza de los Ríos y Loyo (1768–1817), who wrote comedies in addition to hosting a literary *tertulia*, the equivalent of a French salon.[102] She would go on to translate a biography of the physicist Benjamin Thompson, Count Rumford (one of the heroes of economic societies because of his invention of a nutritious and cheap soup for feeding the poor). The Junta's ranks also included María de Betancourt, sister of the famous engineer

Agustín de Betancourt, who, in Tenerife, made trials of textile dyes and helped her brother to develop a twisting machine.[103] And, of course, Amar y Borbón would also be elected a member of the Junta. By 1800 the society could boast of an extended network of about sixty members from the Spanish provinces as well as European embassies. Interestingly, the Marquise of Fuerte-Híjar, the Marquise of Espeja, and the Duchess of Osuna sponsored feminine branches of local economic societies, thus extending the influence of the Junta to other Spanish provinces.[104]

One of the areas in which the Junta soon showed itself to be very active was in the textile industries. Soon after its inception, the Junta commissioned the Marquise of Someruelos to investigate the state of silk manufacturing in Spain.[105] Some evidence suggests that the Marquise of Truillas, together with male members of the Real Sociedad Económica, played a diligent role in textile-dyeing experiments in the plains of the river Manzanares.[106] Between 1788 and 1790, craftsmen and women from Barcelona, Zamora, Córdoba, and Valencia sent cotton, velvet, and other cloth samples to be evaluated by the Junta. They also applied for royal privileges for their business through its mediation.[107] Even some foreign craftsmen who wanted to establish themselves in Spain sought the Junta's recommendation: a female entrepreneur sent a silk spinning machine for them to examine, and the Countess of Superunda tested a spinning machine for three people in 1796.[108]

The Junta was considered by contemporaries as an important achievement of the regime of Carlos III. Sempere y Guarinos, for instance, whom we encountered at the beginning of the chapter praising the warm reception of the *Spectacle* among Spanish gentlewomen, valued the Junta as the first female civic organization in Spain. He stressed the virtue of the Junta's secular character and downplayed the significance of other female societies set up for "exercises of piety and devotion," which is indicative of the changing perception of women's social role.[109]

We do not know if the women who joined the Junta would have preferred to be members of the male Madrid Economic Society and if they were disappointed with the king's resolution setting up a female committee instead. Probably, Amar y Borbón would have felt disheartened. Yet, as Paula Demerson and other historians have shown, the members of the Junta always considered both societies, the Junta and the Madrid Economic Society, to be "branches of the same body," sisters societies, neither of them above the other. Moreover, rather than continue the path of Amar y Borbón in stressing the equality of the sexes, the Junta skillfully used the gender difference as a winning trump card whenever they needed it, for example in their discussions with the Madrid Economic Society or the Royal Academy of Medicine, as we will see in the next chapter. Whether the stress on a feminine nature that made them more suitable for certain enterprises than men was a conscious strategy, the result of an interiorized identity, or the way the Junta construed their public image, is difficult to prove.[110] In all cases, the Junta opened a venue for gentlewomen to enter into politics and ensure that their ideas, as Amar y Borbón put it, would not remain secluded within the walls of the home.

2

Putting Order in the Foundling House

After more than six years of ongoing discussion, the Spanish Crown finally allowed the Junta de damas de honor y mérito to inspect the Madrid Foundling House. On January 24, 1796, a commission of four members scrutinized wards and offices, taking note of all they perceived: the air thick with the smells of rancid fat from the soups, the smoke from the oil candles and charcoal braziers, and the stench of babies' bodily waste; the ragged straw mats on the floor that wet nurses shared with the children; the untidy and dusty filing cabinet and the careless way in which the accounts were kept.[1] Two months later, the Junta submitted a devastating report, which likely played a role in the Crown's decision to hand over the foundling house to the society in 1799.[2] The Junta would replace the institution's former governing committee of clergymen and civil servants and would have total authority over the house: from organizing its everyday life to contracting medical staff, allocating budgets, seeking funds, and authorizing anatomical dissections of deceased children.[3]

A female society comprising the board of a large foundling house in a capital city was an oddity in late eighteenth-century Europe, even for a society as respectable as the Junta. Women were welcomed as benefactors, inspectors, employers, and caregivers, but they did not oversee institutions such as these.[4] In a way, this development reflected the political significance the Junta had attained since its establishment in 1787. Its importance becomes clear when we consider that foundling houses sheltered not only illegitimate children but also those of poor families who could not afford another mouth to feed. A number of broad social issues at the time drew the increased attention of the society to foundling houses. In particular, scholars have emphasized changing attitudes toward poverty, charity, and the duties of the state; the emergence of new policies for increasing the working population; and new views on sexuality, motherhood, and childhood.[5] In this context, the high mortality within foundling houses became especially problematic.[6] The lowest rate appears to have been that of the London Foundling House, with the 43 percent, while others, such as the massive foundling house in St. Petersburg, had reached rates as high as 99 percent.[7] Moreover, the number of abandoned infants had increased over the course of the century. In France alone, forty thousand babies were abandoned annually during the 1770s, an alarming number precipitated by bad harvests and war.[8]

The Junta touted the administration of the foundling house as a break from the previous old-fashioned, careless, and masculine way of running it. Showcasing their knowledge on infant care, household keeping, and homestead administration, the Junta was determined to demonstrate that a learned society of high-ranking women would do better than a commission of clergymen and civil servants. As we saw in the previous chapter, the Junta had been set up almost ten years earlier as the female branch of the Madrid Economic Society.[9] In one way or another, its mostly aristocratic members had demonstrated that they were worthy of membership: some were known for their literary works, others for their ingenious inventions; most led cultural gatherings and had close ties with male reformers; and all were considered learned women.[10]

Yet, overseeing the Madrid Foundling House would prove a challenging task. The house received between eight hundred to twelve hundred abandoned children annually, roughly a quarter of all births in the city, and the mortality rates exceeded 96 percent.[11] The running costs of the house were enormous; the budget had to provide for the salaries of about eight hundred wet nurses and the staff, which included physicians, accountants, laundresses, nurses, and helpers, as well as all the necessary supplies (coal, oil, diapers, bedding, medicines, and food, and even masses for benefactors).[12] As soon as the Junta took command, it instituted reforms, such as doubling the available medical care, refurbishing the wards, instituting new daily hygienic protocols, and controlling the health of wet nurses.[13] However, with no prior experience overseeing a foundling house or, for that matter, any medical expertise, the Junta needed to prove that its initiatives were working. As Ted Porter has argued, numbers give legitimacy to actors who otherwise lack it, since they convey ideas of objectivity, accuracy, and detachment from the producer and the local conditions.[14] Thus, numbers—specifically, mortality rates, the new measures of medical success—were vital for proving the Junta's competence.[15] In order to calculate accurate mortality rates, during the first two years, the Junta revised and revamped the old system of registering, identifying, and tracing children that had been in place since the seventeenth century.[16] Yet in 1802, it definitively changed the old protocol as it seemed that it led to mistakes in the calculations.

In the following, we will see the consequences of these changes, in particular, how they helped to shape the roles high-ranking women could play in improving society and created medical knowledge.[17] Specifically, the ways of registering the entrance and exit of children were similar in all foundling houses at the time and had much in common with mercantile practices. However, slight differences in a protocol could be highly consequential. For instance, at the Coram Foundling Hospital in London, children were routinely renamed as a way to erase their past, giving them the chance to start a new life without the stigma of their origins. In the Hospital degli Innocenti in Florence, illegitimate children were concealed in the pages of the registry books as a result of organizing the indexes by family name. By contrast, the Hôpital des Enfants Trouvés in Paris retained all of the details about each child because families had the right to retrieve their children upon paying a fee.[18] Societal values are reflected in humble paperwork, and in the case of foundling houses, it revealed the ways societies thought about their responsibilities for them.[19] Under the Junta, the new method of registering the children showed how these women conceived of the abandoned infants: as individuals whose biographies had medical relevance and needed to be gathered and known from birth.

The first section that follows shows how the Junta addressed the pressing issues of counting, identifying, and decreasing the mortality of the children under their care. We will then delve into the nitty gritty of the everyday workings of the house in order to fully understand the consequences of changing the traditional system of inscribing its inhabitants. In the conclusion, we discuss how the Junta dealt with the problematic issue of mortality rates.

Counting Babies

In the late 1790s, population policies were at their peak. The wealth of the state was understood to be grounded in its working force, and every measure for increasing it was strongly supported.[20] From the beginning, the Junta underscored their full awareness of the socioeconomic implications of high mortality rates in the foundling house. Yet, its members also always presented themselves as having the sensibility toward children's issues allegedly characteristic of women. Consider, for example, the petition of 1789, the first time the Junta expressed interest in taking over the administration of the house. After mentioning women's maternal qualities (patience and natural affection for children), the secretary of the Junta, the Countess of Montijo, laid out the differentiated political roles of men and women: "Men govern republics with their wisdom and intelligence. The way women can contribute to the public happiness is by taking care of these innocent children."[21] The Junta framed their patriotic aspirations on different grounds than men, as a natural extension of their skills and domestic duties—taking care of young children.[22] This could also be seen in the petition of 1796, in which the then-president, the Countess of Trullas, poignantly invoked the Junta's moral duties to care for abandoned children: "The innocent victims of their parents' disorders, or of their extreme poverty, hold their arms out to us and require our protection, our care."[23]

Notwithstanding, the Junta also showed that they had the knowledge and moral obligations of any other male patriotic body. Moreover, as a society, they placed themselves above the commission of four clergymen and a civil servant that previously governed the foundling house. In the petition of 1789, for instance, along with stressing maternal sensibility, the Junta demonstrated that it was well versed in political arithmetic—that is, the interpretation and manipulation of population data.[24] The Junta compared the infant mortality rates in the foundling houses of Germany, France, England, and Holland with those of Spain, exposing that the mortality rates in the Spanish foundling houses were much higher.[25] Using the "tables of general mortality" (tables that stated the life expectancy of different segments of the population), the Junta calculated the number of lives "lost for the state."[26] According to these tables, although half of children were expected to reach the age of ten, in the Spanish foundling houses, only the 10 percent of children did so. Of this percentage, four-fifths were expected to reach the adult age. The Junta then calculated the loss over 30 years: 76,800 people—an issue that was serious enough to warrant the attention of "any patriotic society," that is, the Junta: "Loss as exorbitant as piteous . . . requires the most serious attention of any patriotic society to find out and bring to an end the causes that produce it."[27] When the Junta finally was allowed to begin their work in the foundling house, they certainly gave the most serious attention to numbers. Along with an agenda that included controlling the transport of foundlings in and out of the house, doubling the medical personnel, ventilating, fumigating, and other hygienic measures, and attending to how children were nourished, the Junta kept a tight rein on the way children and their wet nurses were registered. In fact, the Junta soon discovered that 313 children were missing from the balance sheets.[28] They stated with dismay that this meant that they could not accurately calculate the "issue that most interests us [is] the mortality rates of these innocents."[29]

Yet, knowing where the children were at all times was not without its complications. Babies flowed in and out of the foundling house without ceasing. Each day, three or four children were received; after spending some days in the wards, they were then sent to be wet nursed in the neighborhoods. Only babies too sick to be given away or waiting to be allocated were cared for in the foundling house; their number were around one hundred, and they were attended by twenty to thirty internal wet nurses who lived in the house. But children were often returned and had to be given out again; it was not unusual for a child to be passed around by three or more different wet-nurses. If they survived long enough, children who reached a certain age (eight for boys, seven for girls) were sent out to the orphanage. In addition, some parents came to the house to recover their children once the family situation that led to the initial abandonment was resolved. The numbers for March 1801 are illustrative of the traffic. That month, 115 new babies were received. Ninety-six remained in the wards from the previous month. Sixty-nine were sent out to be breastfed, but forty-two were returned. Four reached the age to be sent to the orphanage, and eleven were given back to their parents. Seventy-four died in the wards and twelve outside.[30] By the end of March, ninety-five children survived in the wards.[31]

As the above account suggests, it was not only a question of counting but also of keeping track of children's identities during all these wanderings. A mistake in identifying a baby could have serious religious, moral, and financial consequences. For instance, the salaries paid to wet nurses depended on the age of children. Before children turned eighteenth months old, they were lactated and wet nurses' salaries were higher. Of course, if the child died, the contract with the wet nurse was cancelled; so, to be paid, wet nurses had periodically to demonstrate that their nurslings were alive, usually by presenting them in the foundling house. According to the report the Junta issued before taking over, children were lost, swapped, and used for begging; dead children were not reported; and wet nurses sometimes defrauded the house, exchanging their dead nurslings with live ones.[32] From checking that children were baptized, to returning the correct child to its family, to paying the salaries of the wet nurses, to calculating mortality rates, it was necessary to be able to assure a child's identity.

Yet, the transient nature of the foundling house caused other medical complications. How could the health of a continuous flow of children be monitored? Similarly, claiming that mortality rates reflected the success of the hygienic and medical measures was not straightforward. What if children entered the house already in poor health? Or what if children died in the hands of external wet nurses?

Mothering Transient Children

The Junta tackled the transient, extended nature of the foundling house in three ways. First, it created a stable female administrative structure, in which it placed every confidence. It contracted nuns from the French order of the Sisters of Charity, who specialized in the care of sick people and children, to live in the house.[33] They became the arm of the Junta, the heart of the house, the everyday mothers of the abandoned children. They controlled the health and behavior of the internal wet nurses, made sure that children were not misplaced, organized the house's daily chores, and kept the house on a budget. Second, the Junta reinforced the paper system for tracing where the children were located at any given time. Finally, it created a network of clergymen and doctors to maintain control over the behavior of external wet nurses and their charges.

The Junta was thus able to organize the foundling house according to the norms of a well-managed household, namely, keeping the accounts, overseeing servants and spaces, and caring for the emotional, moral, and physical welfare of the family. The contract that the Junta made with the nuns was unambiguous as to how they should deal with the foundlings, wet nurses, and accounts. They would earn a salary of forty reals in addition to food and clothes (their habit), but they would be subjected to the orders of the Junta.[34] Every Saturday, the two women directly supervising the foundling house (*curadoras*) sat with the highest-ranking nun to update the accounting books. Clothes, medicines, kitchen expenses, and so on were all carefully written down in the books and reviewed by members of the Junta.[35] During the first months of the Junta's tenure, one member made daily visits at different hours to evaluate what changes should be made. The Junta employed a second clerk for the house's office, a second doctor, and second surgeon.[36] It reallocated spaces according to function and efficiency: it set up a room for dirty clothes, a room for children not baptized, a room for the bodies of dead children, and a room for sick children separate from the rooms for healthy ones.[37] All the cooking was centralized in the kitchen and wet nurses were forbidden from cooking their own meals in the house. Because laundry was thought to be bad for children's health, the Junta outsourced it.

Yet the Junta's primary concern was with the wet nurses on whom the health of the entire house hinged. Reducing infant mortality rates depended first and foremost on devising an effective system to manage the health of wet nurses, as the lives of wet nurses and children were intimately interwoven. If there were not enough wet nurses, suckling babies would starve or become sick from the gruels that were used to ease their hunger (we will discuss this issue in chapter 3); moreover, if a wet nurse or a baby had a contagious disease, the latter would propagate rapidly to the community.[38] In short, the Junta struggled with two kinds of problems. On the one hand, it needed to employ enough healthy wet nurses. On the other, it had to ensure that the wet nurses did their job properly.

Working in a Foundling House

As historians have pointed out, the difficulties of finding enough institutional wet nurses were due in part to rivalry with the private market. Despite an intensifying campaign against "mercenary breastfeeding" in the late eighteenth century, a high number of children were still being sent away to be nursed, which meant that the only wet nurses available for hire were those who could not find a job on the private market.[39] In addition to enabling working mothers, merchants' wives, and helping the upper classes maintain their duties and social lives, other reasons for sending the babies away had to do with notions of the countryside being healthier, the belief that sexual intercourse while lactating was harmful for the milk, and the perception that the reproductive duties of gentlewomen could be halted by lactating. The extent of the wet-nursing phenomenon has been well studied in Paris, where the trade was regulated by the Crown through the Bureau Général des Nourrices; the proportion of children who were wet-nursed seems to have been similar to that in Madrid and other large European cities.[40] Thus, of the twenty thousand babies born each year in Paris at the end of the century, only a few thousand were breastfed at home. Twenty percent were sent to expensive wet nurses on the outskirts of Paris, while half the newborns were placed with rural wet nurses offered by the Bureau. The remaining 20 to 25 percent were abandoned in foundling houses. In Madrid, well-to-do families often sent their babies to be breastfed in nearby villages.[41]

The institutional wet nurses were not only the worst paid but suffered the worst working conditions. Take for instance the case of María Nieves Ruiz. Widowed in early 1800, she decided to employ herself in the foundling house, where she would receive two reals a day for her milk—the equivalent pay of an unskilled male worker.[42] She was expected to breastfeed two children, but if needed she would feed three or even four—yet none of these babies were hers. Her two-month-old son, Silberto, was sent to the nearby village of San Sebastián de los Reyes to be wet-nursed for twenty reals a month (paid by the foundling house).[43] Ruiz might well have considered the health risks that came with the job.[44] According to contemporary physicians, wet nurses were exposed to skin infections, fevers, back pain, and worst of all, syphilis. Although children suspected of syphilis were fed with gruels or animals' milk, symptoms sometimes did not appear until a child was two months of age. Lactating foundlings, often the children of prostitutes or of syphilitic men, were thought to be particularly risky, as syphilis was thought to be transmitted through syphilitic outbreak on the mouth of the baby to the nipple.[45] As one Spanish physician described it, "We see honest and healthy nurses who received this plague from children from the same conduct that gives them life."[46] Ruiz might also have been blamed for being lazy and sloppy, accused of drugging children, using remedies considered superstitious, or of transmitting both physical and moral diseases. For instance, a former rector of the foundling house, Pedro de la Vega, complained about the unruly wet nurses, who, aware that they were indispensable, occasionally mutinied.[47]

The women of the Junta coped with wet nurses similarly to how they dealt with other poor, stigmatized women in the society's other enterprises, such as in the case of the female inmates of Madrid jails.[48] It improved their working conditions, took measures to reform their public image, and attempted to educate them in moral and religious values. They raised the salaries of both internal and external wet nurses. Now, instead of the twelve reals paid to external wet nurses upon weaning, the Junta paid twenty reals. It also increased the control of their health and the quality of their milk, as well as instituting semimonthly inspections of the house by physicians and weekly or more frequent inspections by the nuns.[49] It appears that the Junta also protected wet nurses from the abuses of other employees. In 1800 it fired the surgeon Tomas Mas, apparently because he treated the wet nurses badly.[50] Perhaps the women of the Junta, who agreed with the pedagogical ideas of the French writer Mme de Genlis (see chapter 4), also shared her thinking about wet nurses.[51] Genlis was one of the few voices that sympathized with the women engaged in this profession. Rather than seeing them as bad mothers, as Rousseau had claimed, Genlis argued that abandoning her child to feed others was often the only alternative for a poor woman, because otherwise her child would die. As she put it, this mother "shewed herself to be possessed of real tenderness."[52]

In any case, the Junta transformed the wet nurses' quarters in the foundling house from filthy institutional rooms into spaces that resembled clean, ordered households in which virtue, proper behavior, and arguably maternal sentiments reigned. Under the former, male administration, wet nurses shared a large mattress on the floor with their babies. The Junta installed trim beds with wool mattresses, which were considered easier to clean, and numbered cradles, which were placed on either side of the bed to facilitate the suckling. It improved the stock of basic household items such as dishes and new bedclothes.[53] It ordered that the babies' dirty clothes be removed and the rooms ventilated twice a day and the floors disinfected with vinegar or another "antimephitic," in order to stop the contagion of disease thought to be caused by contaminated air.

The Junta ordered life in the wards down to the smallest detail. Fixed hours were set for lactating, which were regulated by charging the nuns with waking the wet nurses at night to feed (in fact, such stringent protocols led the physician Ignacio Rúíz de Luzuriaga to accuse the Junta of ruling the foundling house like a convent: "Children are fed at bell calls, as in convents," he wrote.)[54] Concerned about the proper behavior of wet nurses, the Junta specified in the contract with the nuns that they should monitor them. Nuns were supposed to forbid squabbles (*riñas*), hubbubs, and "indecent conversations," as well as to assure that wet nurses handled their babies in a particular tender way: "The ward sister will have as her major concern that wet nurses dress the babies during the stipulated hours, with the correct order, tenderness, and silence; correcting them if something was done carelessly or in a hurry, or in case a child was too tightly dressed or badly wrapped, or not properly cleaned."[55] The external wet-nurses, meanwhile, would be controlled through the visits of clergymen and physicians.[56]

The Junta also attempted to increase the attachment of the wet nurses to their nurslings, to forge a motherly tie of sorts. For this purpose, the sister of the ward carried a notebook with her that listed the names of the wet nurses and the crib numbers of their nurslings, with the aim of assuring that the wet nurses always lactated the same children, namely the ones in the cribs beside them.[57] In addition to serve as a precaution against mixing up the children, the measure, according to the Junta, led to a "greater affection" in the wet nurses for their babies. Moreover, as paradoxical as it might seem, the Junta considered that the nuns taught family virtues to wet nurses. In the annual report of the activities of the society, the secretary wrote, "The social and homely virtues [of nuns]—these kinds of virtues which spread gaiety and strengthen the bonds of families, are living lessons that wet nurses and all employees received daily."[58]

We do not know whether wet nurses felt more oppressed under the rule of the Junta than under that of the former male committee. What we do know is that the Junta publicly contrasted its board with its predecessor in unambiguous terms, stressing its feminine—and better—governance of the foundling house. One important public platform the Junta used for this purpose was the annual prize ceremony, in which pupils of the "patriotic schools" (*escuelas patrióticas o populares*) were examined in front of a large audience, including members of the Madrid Economic Society and civil servants.[59] Each year a member of the Junta gave a discourse honoring the queen, *Elogio de la Reyna* (Praise to the queen), which was later printed and luxuriously bound in Moroccan leather and given to the queen, who was one of the principal benefactors of the foundling house.[60] The image of the queen as "the mother of citizens" that these discourses projected not only benefited the Crown but also reinforced the role of upper-class gentlewomen in public welfare. An interesting example from these *elogios* dates from 1801, when the Countess of Castro-Terreño denounced the former male committee, claiming that during their tenure, the wet nurses had "poisoned milk" and their breasts were "full of mortal humor," a reference to syphilis. The wet nurses, she continued, used the children as "dogs" for relieving their breast (a reference to the practice of using puppies to remove excess breast milk, something that was done because it was considered dangerous to women to have excess milk that they could not express).[61] The image of selfish, contagious wet nurses and the former board's negligence and blatant disregard for the welfare of its charges could not have been more different than the picture of loving, healthy wet nurses of the Junta. The Countess of Castro-Terreño also stressed the drop in the mortality rate at the house from 96 percent during the former administration to just 42 percent in 1801 under the "the maternal care" of the Junta.[62]

The same praiseworthy image of a renewed and reformed foundling house was put forward two years earlier by Josefa Díez de la Cortina in the annual prize given in 1799, just two months after the Junta assumed control over the foundling house. She called the old foundling house an "insatiable sepulcher of children" and contrasted this image with its current state: "Oeconomy (*economía*) has followed disorder, intelligence [has followed] ignorance, and an active and protective care [has followed] mercenary negligence . . . Humanity congratulates itself for such an interesting project: the state will have new individuals, and the arts will have new arms."[63] Contrasting the former board's financial motivation with the caring attitude of the newly in-charge women, and stressing the importance that knowledge had in the success of the enterprise (opposing intelligence to ignorance), Díez de la Cortina pointed out the positive results of the reforms of the Junta, an all-female society, for national improvement. But perhaps most important for the sake of the argument is her use of the word "economía."

Oeconomy (as it was spelled at the time in English) is employed here in its eighteenth-century sense of the "prudent stewardship of people and things."[64] The early modern meaning of *oeconomy* did not map onto our meaning of economy.[65] The term originally derived from the Greek and was applied to the governance of the household. The literature on oeconomy included advice on such variegated topics as cookery, cleaning, ordering the spaces of the house, governing the servants, husbandry and agriculture, gardening, and domestic medicine. By the 1780s Spanish writers used both the noun (*economía)* and the adjective (*económico/a*) profusely. They wrote about *sopas económicas* (soups made with cheap and, arguably, nutritious vegetables, conveying this idea of prudent management of resources), *camas económicas* (beds that could be used for different purposes), and of course *sociedades económicas*, *economía política*, and *economía rural*. The adjective could also be applied to physiological process, such as the "oeconomy of digestion" and "animals' oeconomy." In all these cases, they meant the wise matching of form to function, the prudent adequacy of means to objectives, the interrelation of all parts. In fact, as Simon Werrett eloquently showed, to be oeconomic was considered one of the virtues of the well-off, which men and women should cultivate.[66] One should spend what was necessary for decorum, but without wasting.

To learn to be oeconomic was particularly advised in the pedagogical books for women, as a preparation to their role of wives. In his well-known treatise on the education of girls, *Traité de l'éducation des filles*, François de Salignac de La Mothe Fénelon contended that "a prudent woman . . . is the soul of even the biggest of houses: she puts order in the economy, mends the souls, and fortifies the health of the family."[67] Josefa Amar y Borbón (mentioned in chapter 1 as one of the champions of the setting up of the Junta), described the obligations of wives as "the prudent arrangement of rents with necessities" and "the continuous vigilance over personal property and its interest."[68] Moreover, as it was expressed in the statutes, the Junta's principal goal was precisely to propagate this oeconomic behavior among the female population.[69] Hence, Díez Cortina's mention of how oeconomy substituted for disorder, read along with the other allusions to the maternal care of the Junta toward the foundlings, may be understood as a way to underscore how feminine nature, along with the learned skills of educated women, could be of service to the common good.

In the same manner that the control of finances was done through accounting books, the control of how many children thrived or died could be done through children's records books. In the following, we will see the changes that the Junta introduced to better calculate mortality rates, but to do so, it is necessary to delve first into the intricacies of the foundling house bureaucracy.

Babies' Journey through the Books of the Foundling House

The big books for registering the intake of children had three different functions: they worked as church parish books, as catalogues in libraries, and as accounting books in merchant companies. The original purpose of the Catholic foundling houses was to assure that all abandoned children were properly baptized, thereby safeguarding their entrance to heaven (baptism was believed to erase original sin; if children died unbaptized, it was believed they would remain in limbo). Once the clerks of the house checked a foundling's proof of baptism (or if absent, placed them in a special room to be baptized directly), the child was registered in what was called the *Libro de Entradas de Criaturas* (Book of entrances of children).[70] Each year, a new book was opened and numbered, linking each year to a designated number. For instance, the book number 169 corresponded to the year 1801.

The ritual of inscribing a child into these books reflected the way children were inscribed in all church parish books across villages in Catholic Spain. It solemnly and publicly acknowledged the entry of a new member into the Christian community. The clerk of the house chose the first blank page. For centuries, all entries had begun in the same way: "On the [day] of [month], was received in this Holy House, a [boy or a girl] who. . . ." The clerk then wrote down the little information that was known about the child: who had brought the child and where this person lived, the name of the child's mother and father, the date, time, and place of birth, and finally, the parish in which they were baptized. Once registered, the child was officially under the care of the foundling house. Figure 2.1, for example, shows the registration of María Salomé de Baza.

From then on, the information entered into the *Libro de Entradas* would be linked to the actual body of the child: she would be labeled with a little piece of parchment tied tightly to her waist. Foundling houses used the same method that naturalists and librarians used to keep track of their specimens and books—that is, labels. The *pergamino* was a small rectangular piece of parchment of about four centimeters by ten centimeters, similar to the ones shown in figure 2.2. The name of the child was placed in the center of the parchment, the book number in the upper left corner, and the page number

Figure 2.1. Libro de Entradas de Criaturas. Archivo Regional de la Comunidad de Madrid. Fondo Diputación, Inclusa.

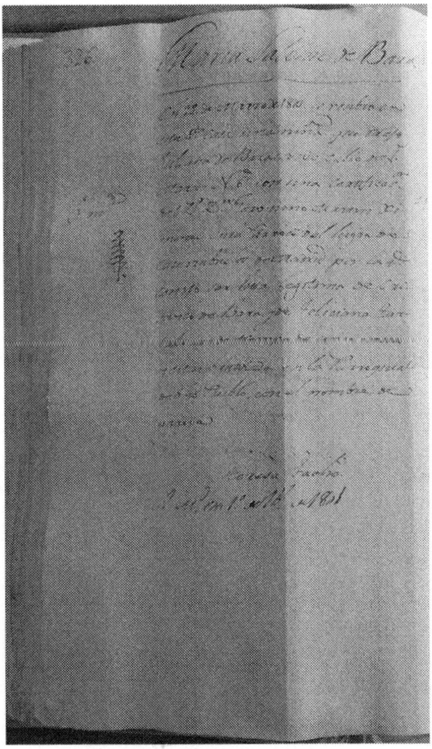

in the lower right. As long as the child was under the care of the institution, these two numbers would be used to identify her (figure 2.2). Thus, the Libro de Entradas not only had a civil and religious role—it also worked as a search tool that linked names, bodies, and numbers (see figure 2.3, children's names ordered alphabetically as they appeared at the beginning of the *Libro de Entradas*). Once the parchment label was tied to a baby's waist, she was ready to circulate through the foundling house.[71] First the baby was sent to the in-house surgeon; after he had checked her over and determined that she was not ill, she was sent to the wards for healthy children. Here, the ward sister noted the baby cradle number in her notebook and assigned her to an internal wet nurse.

Once an external wet nurse was found for the child, a new page was opened in another notebook, the *Libro de Salidas de Criaturas* (Book of exits of children). Similar to the *Libro de Entradas de Criaturas* in size and appearance, the *Libro de Salidas* registered the exits of children who were to be

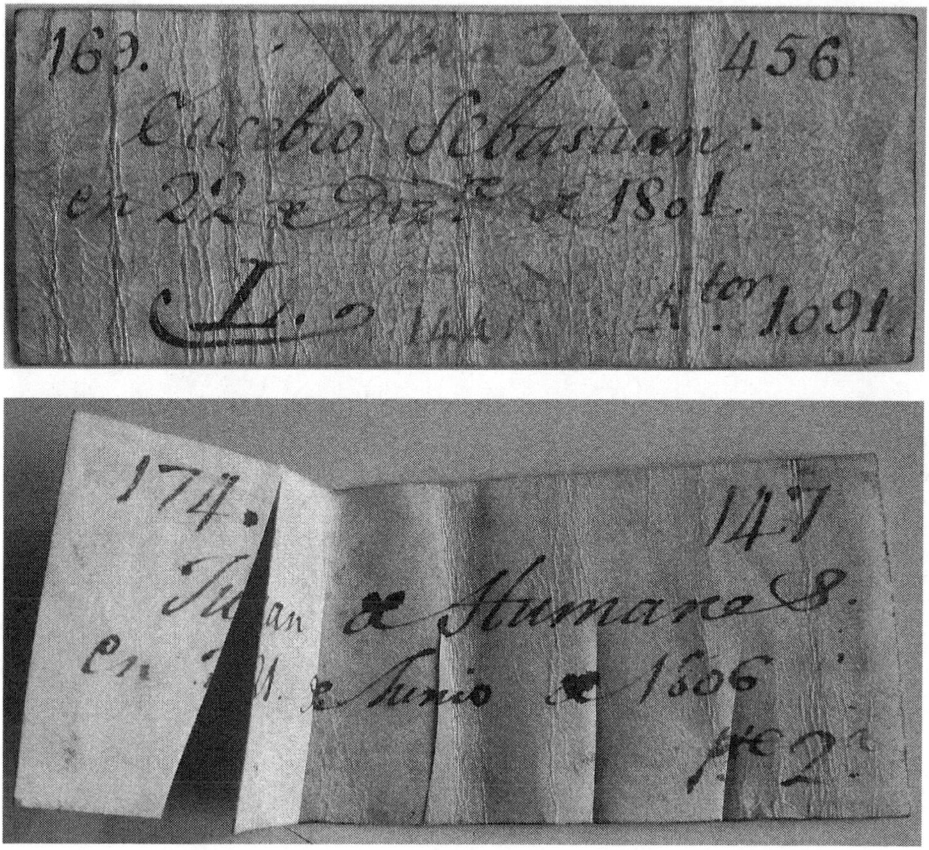

Figure 2.2. Parchments. Archivo Regional de la Comunidad de Madrid. Fondo Diputación, Inclusa.

fed outside the foundling house: the name of the wet-nurse, sometimes the profession of her husband, always her address, and the wages paid to her. In this way, it functioned as an accounting book, but it also kept information on the child's location. For instance, in 1802, one year after nine-month-old María Salomé was abandoned, the family's economic situation improved and the father came to the foundling house to retrieve their daughter, who was successfully located at the home of her wet nurse and brought safely back to him.[72] Before a child was sent out, the clerk annotated the number of his page in the *Libro de Salidas* on the parchment label, in the left corner.[73] Each time a wet nurse came to be paid, she was required to present the parchment label along with the child to prevent cheating, swapping, or children being lost.[74]

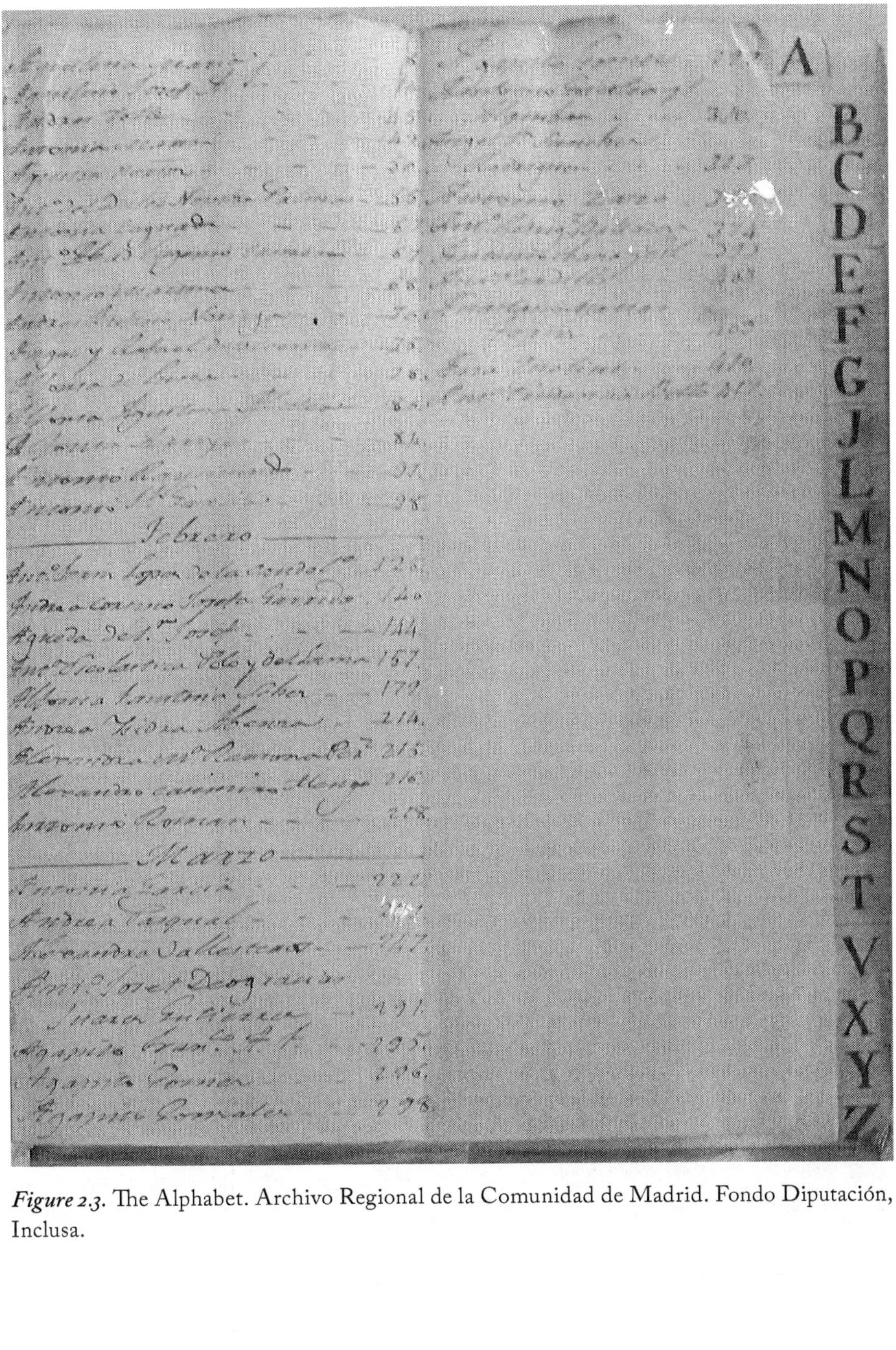
Febrero
Marzo
A B C D E F G J L M N O P Q R S T V X Y Z

Figure 2.3. The Alphabet. Archivo Regional de la Comunidad de Madrid. Fondo Diputación, Inclusa.

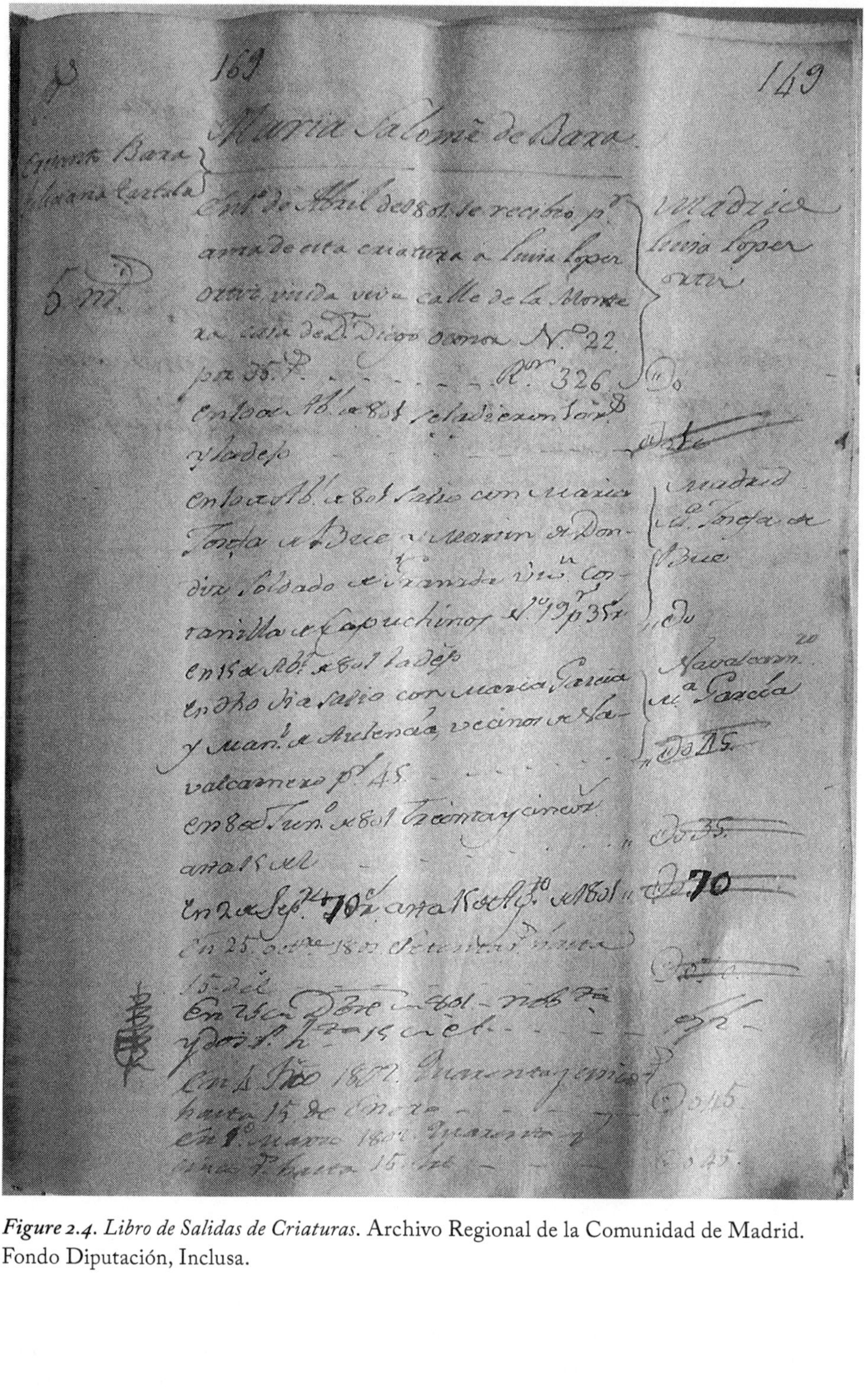

169

149

Maria Salomé de Bara

Figure 2.4. Libro de Salidas de Criaturas. Archivo Regional de la Comunidad de Madrid. Fondo Diputación, Inclusa.

In summary, from a managerial perspective, the Madrid Foundling House was handled like a business. Children circulated from wet nurse to wet nurse like goods traded from hand to hand. Children's register books functioned much as accounting books did for commercial businesses.[75] Consider, for instance, the method of keeping ledger books in eighteenth-century Spanish companies. Manuals recommended opening a single account for each client, in which all related transactions were neatly registered. It was suggested that one write down "all the particulars of each business, all information that would help in handling related matters in the future . . . in a way that in these paragraphs one should find all that is needed for handling credit or debt."[76] In a similar manner, the clerk at the Madrid Foundling House noted all the particularities of the "acquisition" of the child: the date on which it was received, who brought it, where it was found or born, what the names of its parents were, and the expenses it incurred.[77]

Thus, the record books of children played the financial, social, and moral roles that accounting books played in commercial businesses. It was assumed that the way one kept and took care of the books reflected not only one's own moral being but also the respectability of the company. Or to put it differently: accounting books were purposely written for others to look at with inquisitive eyes. In this sense, as Anke te Heesen has put it, "notebooks collected people."[78] In order to avoid deception, formal ceremonies in which the accounting books were revised and signed were staged. In the Madrid Foundling House, a new notebook for registering children was opened and identified by a unique number on an annual basis, and the pages of the book were periodically reviewed and signed by the rector of the institution.[79]

From 1799 until 1802, the Junta worked with the old system of two books (the *Libro de Entradas* and the *Libro de Salidas*). The pages of the books for these years appear clean and well structured. This is particularly evident in the *Libro de Salidas* number 169, from 1801 (see figure 2.4). Furthermore, the Junta encouraged keeping all the information relating to each child, such as notes accompanying abandoned children, certificates of baptism, the parchments of deceased children, and so on, inside the books, tucked into the volume next to the corresponding pages. With their latches and closures, under the administration of the Junta children's record books became more like ordered boxes for keeping precious possessions.[80]

The Junta significantly increased the paperwork carried out in the house. Forms, tables, certificates, and reports were written and filed for ascertaining that children were well tended. Parish priests were enlisted to keep an eye on women already serving as wet nurses in their parishes, while those who wanted to be employed by the foundling house needed certificates of good behavior.[81] Measures for stopping the illegal traffic of parchments were issued.[82] During these years, the mortality rates of the children at the foundling house were calculated using the two books of entries and exits. The Junta sent monthly reports to the Crown; these, over time, showed a substantial decline in the mortality rate. Journals and medical publications also echoed this drop.

However, the Junta was concerned that keeping information about the children in two separate books might complicate the production of statistics as a source of possible error. Finally, in 1802, the society decided to create one unique book for the registration records, the *Libro de Entrada y Salida de Criaturas* (Book of entries and exits of children). Book number 169 was the last edition of the old two-book system. This innovation was publicly announced by the president of the Junta, the Duchess of Osuna, in her annual report: "The two Books of Entrances and Exits have been reduced to one, with a noticeable increase in the order and the clarity [of the page] and with the advantage of putting together what is said of every individual."[83]

The new books optimistically reserved three pages for each child, which, if completed, would encompass a sort of schematic biography. The Junta provided a page model for the official clerks that specified how the space on each page should be organized. The page was divided into two parts by a line: the upper part comprised information about the child, while the bottom described her stay with the external wet nurses. The page was spatially divided for the different functions that it would perform. The creases for dividing it in four equal quarters (two central for the information, two lateral for signatures and accounting) are still visible. Each piece of information had its place, which made it easier to look up the data for any given child.[84]

The tidy, neat, new books were the material representation of the new order that the Junta imposed; they represented its authority and governing capacity, demonstrating how a society of learned women—applying their maternal and economic skills—could drastically diminish the mortality rates and produce reliable statistics of service to doctors and political economists.

With the new books, the Junta embarked on more sophisticated statistics. It annotated and quantified the most frequent causes of death and how mortality rates were distributed by age and sex. In 1811 the general procedure of writing down the cause of death allowed for the tabulation of the most frequently occurring life-threatening illnesses, ordered by age and sex. From 1802 onward, we have records of a new notebook, the *Libro de Muertos* (Book of the deaths of children), which had a long and narrow format particularly suited for writing down lists.[85] The clerk made monthly summaries of deaths by age and sex and noted if they occurred in or out of the foundling house (figure 2.5).

But then, how should one interpret the mortality rates? These were supposed to assist in the evaluation of the effectiveness of the Junta's administration, by giving a numerical measure of its success. What if—as it happened in 1805 and 1806—famine and war brought in a crowd of malnourished children who all perished? Would it be the fault of the Junta?[86] As Ted Porter notes, once numbers were used to assess the effectiveness of the institutions, administrators began to do some "creative" accounting, and new ways of interpreting the numbers appeared.[87] Thus, the Junta introduced new categories for data collection of statistical interest. In particular, the new category of "inconservable" (literally: "impossible to keep alive") required scrutinizing the condition in which children arrived, not least to absolve the foundling house of responsibility should they die. Further rubrics distinguished between illegitimate and legitimate children, recording those who came from poor families as well as those who died in the wards of the foundling house or with an external wet nurse, and if the children had been born in the hospital for poor women or at home. The new books thus allowed for a more in-depth assessment of the children's condition, allowing the Junta to take account of their health when they entered and the circumstances of their deaths.

It is tempting to say that the conception of the child changed, in that it was no longer perceived as a commodity that moved from one point to another, as was the case when their data were registered in two books. In the new books, records became compressed stories—as the Duchess of Osuna put it, "all [that] was to be said about each individual" was there. No longer were children's lives split between two books, but instead all the essential details were kept together in one limited and fixed space. At least on paper, the children had a home. Although few children filled a full three pages in

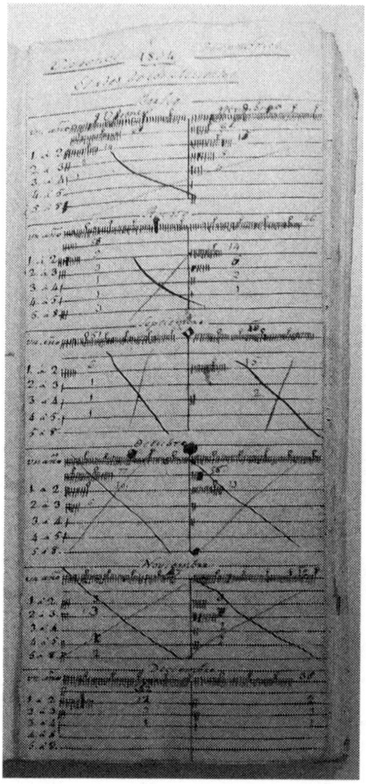

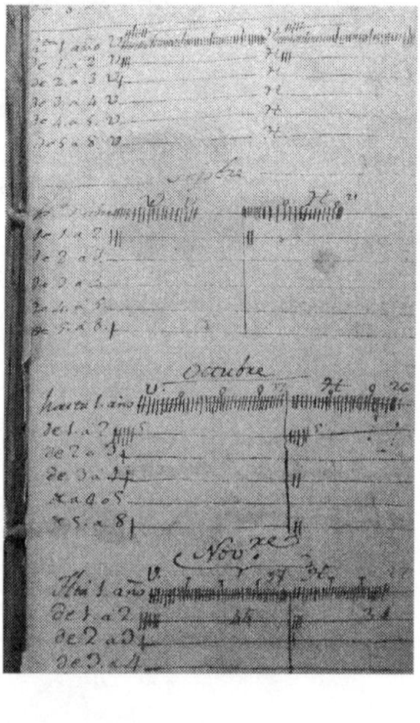

Figure 2.5. Libro de Muertos. Archivo Regional de la Comunidad de Madrid. Fondo Diputación, Inclusa.

the new books, this amount of space was allotted to each in the hopes they would survive long enough to need it.

Was It for the Good?

The Junta construed its tenure as feminine and opposed to masculine and mercenary caring, showing the importance of a society of female improvers in the public space. It shaped its political role as an extension of women's caring role. It transformed a space ruled by clergymen and civil servants into a feminine domain. The Junta organized these spaces, regulated the tasks of all employees, checked accounting books to the last details, and set up lactating wards that enabled emotional bonds between foundlings and wet nurses.

At the same time, the newly ordered and clarified record books were the material representation of the organized, economic skills of learned women, who were used to household accounting and state management. But above all, the books for registering children provided a means to accurately calculate an "objective" measure of success. Numbers can be used as measures of achievement, as rhetorical arguments for demonstrating success or failure. The closer monitoring of children's health also produced hundreds of statistics classifying deceased children in accordance with their age, sex, place of demise, and possible causes of death. If we apply Francesca Bray's definition of gynotechnique—as a "a set of technological attributes and practices that combine to shape and define female roles and hierarchies"—to this case, the activities of the Junta could be considered as a case of gynotechnique applied to the common good.[88]

We can thus ask: was the Junta's new regimen beneficial for the poor classes? On the one hand, the governance of the Junta in the foundling house, as was only to be expected, buttressed the hierarchies of the ancien régime, reinforcing arguments for the usefulness of higher classes in post-revolutionary times. The governance of the foundling house reproduced the social order, but with a qualification: it was governed by women. Doctors and surgeons, suppliers and official clerks, were subject to the orders of the Junta. Lower-class women breastfed the children, nuns took care of the everyday chores, and gentlewomen provided resources and leadership. Nuns and ladies educated the lower classes in religious and social values, monitored their health and the health of the children, and showed them the proper way of mothering.

Although the surveillance of wet nurses increased, it also likely benefited them, as the Junta bettered their salaries, working conditions, and public image. It probably benefited the surrounding population of the foundling house, as the number of jobs offered by the house increased. It certainly improved children's life in and outside the wards and their chances of surviving increased from almost certain death. However, the mortality rates still were still very high, especially in times of famine, when the rate at which children were abandoned increased, and in summer, when poor mothers left their children in the house to work in the fields and there were not enough wet nurses.

The story of how the Junta tried to develop safe alternatives to breast-feeding is the theme of the next chapter.

3

Testing Baby Food

In 1800 the secretary of the Junta de damas, the Countess of Montijo, described an ingenious artefact that allowed babies of the Madrid Foundling House to feed themselves, on their own, from goats' udders: "Because of the abundance of children and scarcity of wet nurses, it was necessary to make use of goats placed on a rack with a matrass underneath, where two babies comfortably suckled from one animal, without having experienced any mishap."[1] The device would be later depicted in a manual on how to rear foundlings published by the physician of the house, Santiago García, and it certainly looked like a kind of living feeding bottle (figure 3.1).[2] The front half of the goat was attached to a wooden frame by a belt that circled its neck. In the rear half, a wider belt with three holes kept the goat's legs off the babies and fixed the udders in place. These were covered by two leather hoods with little holes to control the expression of milk.

The goat-bottle spoke volumes about the Junta's struggle to cope with hungry children and the lack of human resources. When this female branch of the Madrid Economic Society took over the administration of the foundling house in 1799, it intended to end the high mortality of children by doubling the medical staff, establishing hygienic practices, and controlling the health of wet nurses. However, the most problematic issue remained:

Figure 3.1. Devices for nursing foundlings supposedly used in the Madrid Foundling House. In Santiago García, *Instituciones sobre la crianza física de los niños expósitos* (Madrid: Imprenta de Vega y Compañía, 1805). Biblioteca Nacional de España.

how to feed young babies when there were not enough wet nurses. To be sure, the use of alternatives to breast-feeding was customary in foundling houses. Week in and week out, and especially in times of famine when poor children crowded into the house, wet nurses did not suffice. At the same time, paps and animal milk were mandatory in the case of babies infected with allegedly contagious diseases such as syphilis.[3] In the Madrid Foundling House, for instance, pewter jars similar to the one in the figure 3.1 were also used to feed the babies. Easier to clean than horns, bladders, spoons, or cloth wrappings, the jars had a long pipe attached near their base, making the liquid flow slowly to prevent the baby from drowning.[4]

These artefacts, the literature, the iconography, and hundreds of recipes provide evidence that eighteenth-century babies were fed with animal milk, paps, gruels, and broths—what was called "artificial feeding." Yet none of these methods seemed to work in the foundling house. Children often got sick and perished.[5] This chapter focuses on the trials that the Junta conducted in an effort to set up a safe form of substitute breast-feeding and decrease the mortality rate. The information we have about the trials is scanty and scattered in different archives, but it clearly shows that, allied with the physicians of the foundling house, the Junta experimented with goat's milk, donkey's milk, and the flour of a tropical root in 1800, 1803, 1804, 1805, 1812, and 1817.[6] At their core, the trials consisted of feeding babies of different ages with goat or donkey's milk and tweaking doses, dilutions, additions of spices, frequency, and ways of administration to find a method that would work. For instance, how should a week-old baby have its milk, diluted three or two times with water? How much sugar should be added in the pap of a three-month old baby?

We have known about these trials since the pioneering study of Paula Demerson on the Countess of Montijo, who was the long-time secretary of the Junta and *curadora* (responsible for) of the Madrid Foundling House during the first years of the Junta administration.[7] However, how these trials were actually performed, which variables were tested, how and under which logic, and how the relationships of the Junta with the doctors inside and outside the house developed are issues that have not been addressed yet. On the last topic, the existing narrative hangs on Joan Sherwood's thoroughly social study of the foundling house.[8] In it, Sherwood portrays the trials as personal projects of the doctors of the house, and contrasts the "maternal behavior" of the Junta with the "scientific" approach of doctors.[9] For Sherwood, the

Junta represented the feminine way to approach caring, "maternalism in action," as she put it.[10] Sherwood described the Madrid Foundling House as "a human laboratory for new techniques and experiments." In her telling, when the Junta intervened in this laboratory, they did so to counteract the excesses of "the men of science."[11]

By contrast, this chapter acknowledges the agency of the Junta and shows how researching new methods was in fact part of their duty as a society of improvers. Framing the tests in the contemporaneous culture of testing medical treatments in human bodies, researching new food for the poor, and gender expertise, it reveals how the Junta effectively tested, observed, kept diaries, and discussed the results of the trials among themselves. To begin with, the method of trying food in babies did not differ in principle from how people were accustomed to trying drugs and cures. As Elaine Leong and Alisha Rankin have shown, drugs and other cures were tested in the bodies of families, friends, and acquittances, in prisoners, poor people, and animals.[12] These trials meant not only substituting ingredients for others that might be easier or cheaper to obtain but, as Leong demonstrates in her work on early modern seventeenth-century English households, also implied trying doses, methods of application, and techniques of production and preservation. It required observing, recording, and exchanging information with other practitioners.[13] Yet the trials in the Madrid Foundling House differed from that of members of a household tweaking cures in the bodies of their kin, in particular, in relation to the number of individuals involved. Although the scale was not comparable to other contemporary clinical tests, the foundling house experiments still implicated numerous infant and observers.[14] The Junta employed what we today call "control groups," that is, groups of infants who were breast-fed and whose development was compared—according to criteria that differed in different situations—with that of children fed by artificial means. The trials also involved numerous witnesses—the women of the Junta, the physicians, and even the nuns who took daily care of the children.

On the other hand, finding new ways to feed the needy was a major goal of late eighteenth-century alimentary experts, a loose collective "in construction" as Emma Spary puts it, which involved apothecaries, agronomists, gastronomes, food entrepreneurs, chemists, and philanthropists.[15] In a way, seeking alternatives to breast-feeding aligned with the attempts to find a nutritious and cheap soup to feed the poor; with the increasing status of potatoes in statemen's minds as a solution to famines; and with the trials for making edible bread from alternative crops and gelatins from bones.[16] These are but a few examples of how the trials on foodstuff gained importance during the Enlightenment in tune with ideas that related the wealth of a country to the health and vigor of its working force.[17] A female society of improvers, of enlightened women at the service of the public welfare as the Junta regarded themselves, could not but be interested in this kind of research, which if successful would save the lives of thousands of foundlings.

Gender issues also came into play. Women, as mothers, were expected to know the different ways to feed their children and to monitor their health. Artificial feeding along with wet-nursing was considered one of the main causes of infant mortality, so providing guides on how to bring up healthy future citizens aligned with the expectations of a female economic society. However, the feeding of children in general, and foundlings in particular, was also a medical and public issue. The Junta thus needed to negotiate its legitimacy with physicians, not so much with those employed at the foundling home (because these were under its orders) as with the supposed public health experts at the Royal Academy of Medicine of Madrid. I will suggest that it was a two-way relationship, in which the elitist members of the Junta could also influence the careers and prestige of the academicians. The delimitation of each side's responsibility was done through a gendered discourse, which allocated each sex a different sphere of activity.

The story unfolds in four parts. The first situates the Junta in the context of contemporary debates on baby feeding; the second discusses the problematic nature of milk in the late eighteenth century in order to help understand what variables were tested in the trials and why. With these insights as background, we then explore the long trial with goat and donkey's milk carried out from 1805 to 1806 and examine how the Junta carved out their public image. In the fourth section, we move forward ten years, to 1817, when the tropical root *nuevo sagú*, obtained from Cuba by one of the members of the Junta, was tested in eight children. We finish in 1828, with the later reflections on the trials by the former president of the Junta, the now elderly Duchess of Benavente.

How to Feed a Baby in the Eighteenth Century

Before the Junta embarked on the trials described here, there was already empirical evidence that artificial feeding worked in some cases. The general recommendation was maternal breastfeeding, and the vehemence with which the campaign against wet-nursing unfolded in the late eighteenth century is well-known, as was the resistance of a great part of the population to following these recommendations (see previous chapter). Scholars have also illuminated how the reasons for this vehemence were to a significant degree political.[18] Yet, journals, medical treatises, and private correspondents discussed, in detail, cases of children raised with paps; the milk of cows, donkeys, ewes, and goats; and with different "waters," that is, herbal teas that could also contain milk. For instance, the Spanish doctor Pascual Mora registered the case of a widow in London, Mrs. Pollard, who reared seven children with a pap of water and bread, to which she added cinnamon and sugar, and the case of Mr. Williams in London, a widower who fed his daughter with biscuits dissolved in water until she was four months old, at which point he introduced cow's milk.[19]

Moreover, the great emotional value that was associated with feeding one's own child, as well as the fierce campaigns against wet-nursing, fostered the development of alternatives available to mothers worried about breastfeeding their children. Silver nipple shields, suckling glasses, and devices such as the breast pumps produced by the Parisian instrumentalist Jacques Bianchi or the surgeon William Rowley in London were meant to help with problematic breastfeeding; at the same time, some women turned to baby food recipes.[20] In fact, there was a wealth of baby food recipes to choose from.[21] One well-known case is that of Madame Roland (Jeanne-Marie Philippon). She recorded her anxiety when she began to produce less milk and explained in detail how she fed her eight-week-old baby Eudora with bread dipped in sweetened rice water, as well as with cow's milk diluted with water in which barley had been boiled and sweetened with honey, until her milk returned again after seven weeks.[22]

The Junta was well aware of the different methods of feeding babies long before they took over the Madrid Foundling House in 1799. Consider the case of a letter written by the Marquise of Llano (figure 3.2), admitted to the Junta soon after moving to Vienna with her husband, the Spanish ambassador to the Hapsburg court.[23] As a member abroad, the marquise was expected to apprise the Junta of relevant news, so in December 1788,

► *Figure 3.2.* Portrait of Isabel de Parreño y Arce, Marquise de Llano (c. 1775) by Anton Raphael Mengs. Real Academia de San Fernando. Public domain.

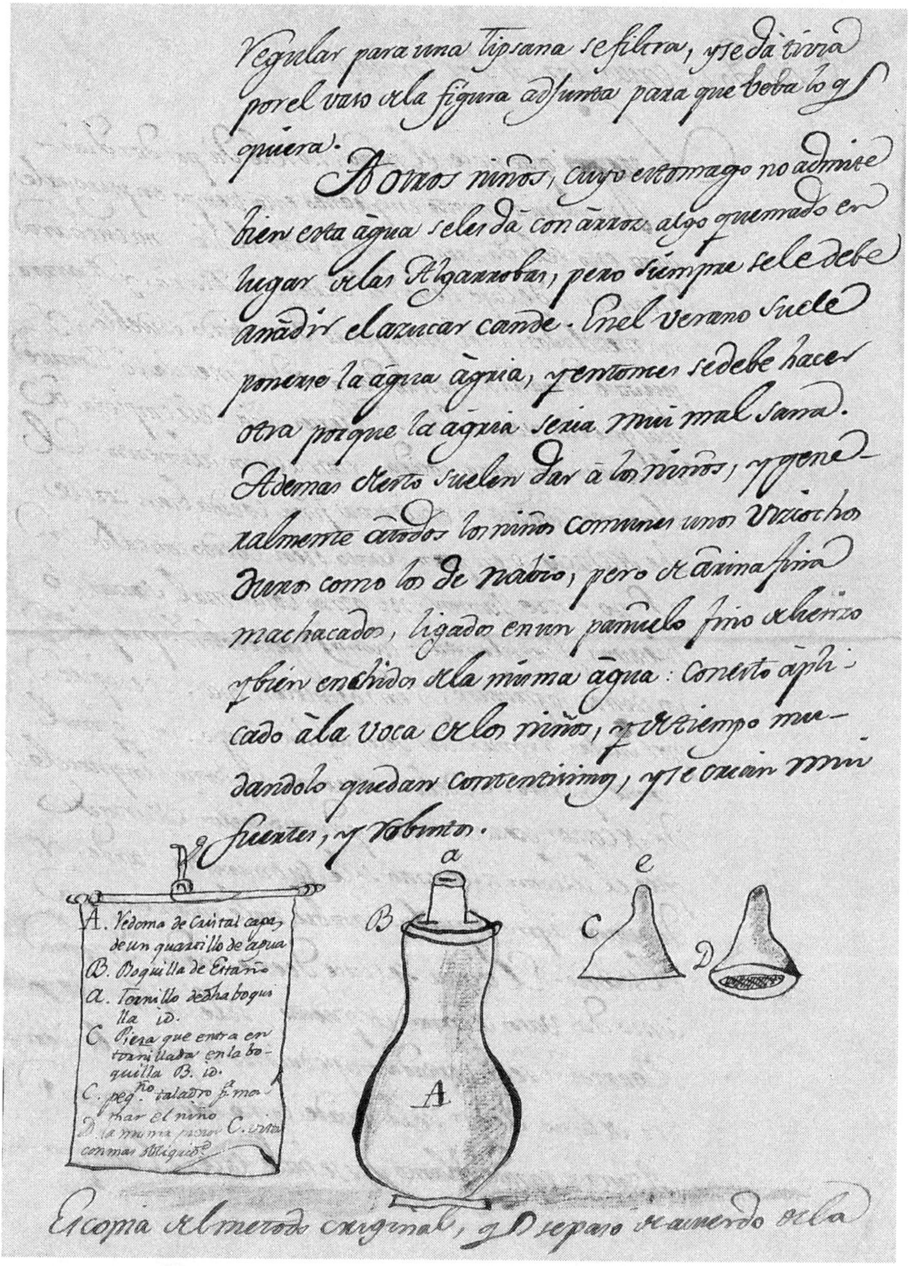
Regular para una tipsana se filtra, y se da tirna por el vaso de la figura adjunta para que beba lo q. quiera.

A otros niños, cuyo estomago no admite bien esta agua se les da con arroz algo quemado en lugar de las Algarrobas, pero siempre se le debe añadir el azucar cande. En el verano suele ponerse la agua agria, y entonces se debe hacer otra porque la agria seria mui mal sana. Ademas de esto suelen dar a los niños, y generalmente a todos los niños comunes unos vizcochos duros como los de navio, pero de harina fina machacados, ligados en un pañuelo fino de lienzo y bien enchidos de la misma agua: con esto aplicado a la boca de los niños, y de tiempo en tiempo mudandolo quedan contentisimos, y se crian mui fuertes, y robustos.

A. Redoma de Cristal capaz de un quartillo de agua
B. Boquilla de Estaño
a. Tornillo de dha boquilla id.
C. Pieza que entra entornillada en la boquilla B. id.
C. peq.no taladro p.a mamar el niño
D. la misma pieza C. vista con mas distincion

a
B
A
e
C
D

Es copia del metodo original, q. D. separo de acuerdo de la

Figure 3.3. Letter written by the Marquise of Llano from Vienna, in which she included a recipe for feeding babies and a sketch of a feeding bottle. Archivo de la Real Sociedad Económica Matritense de Amigos del País.

she put pen to paper and reported on the Austrian textile industry and on baby food. In the letter she sent to the Junta, she included two recipes for bleaching muslins and an infant food recipe, "A Method for Rearing Children with Water." She also sketched a glass feeding bottle (figure 3.3). According to the marquise, this "water" was used in the Vienna Foundling House, but also in the households of well-off ladies who apparently preferred it to wet-nursing.[24] We will discuss this recipe in detail in the next section; here, I want mainly to stress how baby food was a central topic in the context of the Junta's patriotic interests. As a society of women improvers, the Junta was expected to be well aware of the different methods of rearing children, the literature on this topic, and the problems of feeding babies with animal milk and other substitutes.

In fact, the Junta extensively discussed and wrote on the topic of infant feeding. In 1790 Josefa Amar y Borbón—whom we met in chapter 1 in relation to her defense of women's intellectual capacities—published an updated treatise on the education of young women, *Discurso sobre la educación física y moral de las mujeres* (Discourse on the physical and moral education of women).[25] At the time, education was widely considered to consist of both "physical" and "moral" education—"physical education" pertaining to maintaining babies' health and "moral education" being what we would today consider properly education. In the section devoted to physical education, she specifically discussed ways of feeding babies. She limited her considerations to breastfeeding, preferably by the mother, because she believed this was the only way a very young baby would survive; if the mother was not able to lactate, wet-nursing was to be chosen, not artificial feeding. Food was only to be introduced when the baby was around seven months old. Evidence that the book was well received by doctors may be seen in the fact that Amar y Borbón was proposed as an honorary fellow of the Barcelona Royal Academy of Medicine.[26]

Later in 1795, the Junta launched what we could call a research project on girls' education.[27] As in other patriotic societies, the research was organized around questions to be answered by various commissions. In particular, the commission on physical education was expected to address issues related to an infant's proper diet, such as: Should children be given food other than milk? If so, what would be most appropriate? How long is nursing suitable? What is the most appropriate food to help children grow up to be robust? At what age should one begin to give children solid food, what should that food be, and in what quantity? The Junta produced seven memoirs, authored by Josefa Amar y Borbón, the Countess of Truillas, the Duchess of Ariza, the Marquise of Fuerte-Híjar, and the Marquise of Sonora (the Marquise of Sonora would be directly responsible for the Madrid Foundling House during the trials of 1805).[28]

Furthermore, before they themselves embarked on the trials in the Madrid Foundling House, the Junta was probably aware of trials in other cities, as other societies also had the same pressing problem of finding safer substitutions to wet-nursing. For instance, during the 1770s and the 1780s, the Académie Royale des Belles-Lettres, Sciences et Arts of Bordeaux offered a prize to whoever provided "the best and most economical method for providing a substitute to breast milk."[29] The prize was announced several times, in 1778, 1781, 1783, and 1785 and garnered twenty-eight responses. In 1783 the doctor Simone-Victor Lamothe (1736–1823) was asked to test a particular method with the foundlings of the Hôpital des enfants trouvés in Bordeaux, but unfortunately, he had no success. The trials were carried out during the summer months and babies were fed with milk, broth, and sugar. Two died in the week of August 17, the rest (no mention of the total number) were "very thin" and only one "well-enough."[30] In the London Foundling Hospital, the well-respected Stephen Hales proposed an assessment of artificial feeding. The hospitals of Béziers, Douai, Grenoble, Toulouse, and Montpellier also tried animal milk.[31] Due to the relationship the Junta had with reformist elites and doctors, it is plausible that they were aware of trials in other European foundling houses.

Finding safe artificial feeding methods was also on the agenda of other Spanish female societies at work in foundling houses. One example is the Society of Ladies of Málaga (Asociación de Señoras de Málaga), which was set up in 1796 to aid the Málaga Foundling House. On the last page of its statutes, the society stated that it would form "a plan for the physical education of the foundlings" ("proyecto de crianza física de los niños"). The society aimed to elucidate "whether children could be nursed with goat's milk, cow's milk, or 'artificial breasts,'" because "many things have been written and discussed about this in all times."[32] The same task was assigned to the Society of Ladies of Córdoba. In 1798 the Spanish physician Juan Antonio Trespalacios y Mier encouraged the ladies of Córdoba to join a female society for tending to the foundlings of the city.[33] One of the reasons he offered was precisely the capacity of women for assessing whether animal's milk was feasible for infant nourishment: "because in truth . . . they are best at distinguishing if the milk of goats and others animals used in other countries are useful or if they are as harmful as they seem to be."[34] This quotation illustrates well the muddy waters in which the Junta would maneuver. On the one hand, it seems that in many cases and in other countries artificial feeding worked, but on the other hand, there was evidence of its harmful effects. But what was exactly to be tested and how? What were the epistemological assumptions in the trials?

Organic Fluids and Growing Bodies

The tests basically assessed the reciprocal influence of two changing bodies: the milk, whose properties were altered by the lifestyle of the living being that produced it, and the infant. The physical properties of milk, on the one hand, depended on the type of animal—for instance, donkey's milk was thought to be sweeter and lighter than goat's milk and more similar to women's milk. These properties were also thought to change with animals' diet, habits, and environmental conditions, such as whether they exercised or not, breathed fresh air, and fed on good pasture; were healthy; and even had a peaceful existence.[35] The fast-growing bodies of infants change, on the other hand, in terms of their organic needs and their digestive capacity.

The tests looked to three variables to understand why a feeding method sometimes worked while other times it caused sickness and death. The first variable was the age of the baby. Consider for instance the recipe sent by the Marquise of Llano from Austria, mentioned in the previous section. Her *Method for Rearing Children with Water* was based on an herbal tea prepared with carob bean and nib sugar (*azúcar cande*), but the consistency of the water and the frequency of administration changed as the child grew. The first two days, the baby was fed only with two or four spoons of this water. During these two days, the artificial method was supposed to mimic the already recognized purgative properties of mothers' first milk, the colostrum. The water was then progressively mixed with cow milk and biscuits, with the consistency slowly increasing as the child grew to avoid "indigestion." Frequency was also to be considered. Babies less than a week old were supposed to be fed only twice a day (they supposedly needed less food and more rest), while babies older than that were supposed to be fed more often. However, this was also controversial, as some physicians believed that smaller babies should be fed smaller quantities and more often than older ones.[36]

The second issue that needed to be tested was how milks and paps would be administered, as it was thought that this might affect their nutritional and organic properties. Milk was regarded a very unstable substance, liable to losing its gaseous "primordial quality"— or vital properties—in contact with the air.[37] The French physician Alphonse Le Roy, for instance, reported achieving a dramatic drop in the mortality rates of the foundling house of Aix by allowing children to suckle milk's goat from the udder, so the vital gas was not lost. Apparently, goats were trained to enter the nursing room, remove the cover of the babies lying in their cribs with their horns, and straddle the crib to suckle the infant.[38] Other reasons for preferring one or another method was the facility of cleaning, as we saw in the case of the Madrid Foundling House, where metal feeding bottles were preferred over horns and other devices. Another issue was how animal milk should be heated. The famous English physician William Cadogan was against boiling cow's milk, while Santiago García in the Madrid Foundling House gave careful instructions for warming goat's milk by just pouring in warm water to avoid curdling.[39] Perhaps the most radical view against any kind of milk in infants' diet was expressed by Johan Baptista Van Helmont, who advocated for a light pap made from bread boiled with beer or wine, to which sugar could sometimes be added. Although his ideas were already less influential by the late eighteenth century, milk's instability, even inside animals' bodies, was a shared anxiety.

Finally, in addition to doses and frequency and ways of feeding the babies and preparing the milk, the third variable to be tested was the diet of the relevant animals. By the time the Junta embarked on the trials, a knowledge of how animals' diets affected the digestibility of their milk had arguably already been articulated chemically. As Barbara Oland shows, during the eighteenth century, milk became the target of chemical analysis.[40] This is evidenced in the publications of well-known chemists, such as Pierre Joseph Macquer, and by the prizes awarded by the Royal Society of Medicine in Paris (1787) and the Royal Medical Society of Edinburgh (1785) on the properties of milk from humans, cows, goats, donkeys, ewes, and camels. The French prize was won by the well-known savants Antoine-Agustin Parmentier (1737–1813) and Nicolas Deyeux (1753–1837), while the Medical

Society of Edinburgh prize was awarded to Samuel Ferris (1760–1831). Ferris in particular distinguished the yellow milk produced by cattle in summer, when cows were fed on grass, from the bitter and fatty milk produced in winter, when they ate hay and straw. Ferris quoted Thomas Young, Friedrich Hoffmann, Albrecht von Haller, and Robert Boyle, who provided evidence of different colors, tastes, and properties of milk depending on what the animal was eating.[41] Some physicians believed that the milk from herbivorous animals (goats, donkeys, and cows) was fundamentally different from women's milk because of their different diets. Vegetables rendered milk acidic, while meat rendered it alkaline, so some doctors recommended milk from pigs, which are omnivores, as possibly being more similar to women's milk. The milk of goats and donkeys was often diluted with meat broth to add meat's alleged alkaline properties.

From 1800 onward, the Junta along with the successive doctors of the Madrid Foundling House (Santiago García, Antonio Anento, Tomás García Suelto, and Pascual Mora) tested substitutes for maternal food, including animal's milk and the roots of *nuevo sagú*.[42] From some of these tests, we have very little information. For instance, we only know that in January of 1800 the Countess of Truillas and the Marquise of Sonora hired two women to try goat's milk and that in 1803, the Junta rented the house of the Duke of Granada to "feed the children with goat's milk."[43] However, we have enough material to reconstruct how the different variables were tested with the milk of goats and donkeys in 1805, and how the troubled relationship of the Junta with the Royal Academy of Medicine might have unfolded.

Negotiating Legitimacy: The Trials with Animal Milk

On February 18, 1805, during the Junta's weekly Friday meeting, its president, the Duchess of Benavente, announced that the "comparative trials" (*ensayos comparativos*) with goat's milk would begin the following Monday.[44] Six children would be put with wet nurses while another six would be nursed by goats. The president invited the members of the Junta to join the doctors and the nuns of the house to witness the trials.[45] At the next Friday meeting, on February 25, the president confirmed that the tests had "good prospects." This is the last trace of the trials that appeared in the minutes of the Junta. We can recover the thread, however, in the dismaying report that the physician of the foundling house Antonio Anento wrote years later, from which we know that the tests lasted more than a year and that many more children than the initial dozen participated.[46]

According to Anento, the trial involved infants of "all ages," from newborns to eighteenth-month-old toddlers.[47] The changes in their health were dutifully recorded in diaries. These diaries were later sent to the Royal Academy of Medicine and today are lost, but in the cover letter to the academy, the secretary of the Junta provides a clue about how the tests were prepared: "Among the varied measures that the Junta took was to carry out a trial [*prueba*] to rear some foundlings with wet nurses and others with goats according to the plan established by the physician Santiago García in his treatise on the physical rearing of foundlings approved by the Royal Academy."[48] This was García's *Instituciones sobre la crianza física de los niños expósitos,* which included a series of tables on artificial feeding (see figure 3.4).[49]

We can safely suggest, then, that the Junta proceeded in the following manner. Children were divided into six groups according to age. Each group was then divided in two, one fed with goat's milk and the other breastfed. Goat's milk was diluted with water or broth and mixed with sugar and other condiments (cinnamon, herbs, and bread). Older children were fed more frequently than younger ones, and with a thicker pap.

For instance, babies in group two (a week old at most) were given a cup with milk mixed with three parts of water and sugar, three times a day. Group three (one month old at most) was given a cup of milk diluted by half with a "tenue" meat broth or warm water with sugar, four times a day. Group four (two months old at most) was fed with pure goat's milk (either in a cup or directly from goats' udders) six times a day, while group five (more than two months to teething) was allowed, in addition to the six cups of milk, or to suckling directly from the udder, to have two more cups of soup with bread. Almost none of the children who were fed with goat's milk survived, symptoms ranging from diarrhea, vomiting, stool that was "green, yellow mixed with cheesy milk," to convulsions.[50] Sometimes sick infants were given back to wet nurses and recovered. In those who died, dissections revealed a "kind of cheesy mass" in the stomach.[51] On June 28, 1805, almost six months after the trials began, the Junta sent the "very exact diaries" and the results of anatomical dissections practiced on some children to the Royal Academy of Medicine of Madrid.[52]

Here, I would like to pause to consider the strategy of the Junta. The fact that it sent the diaries to this prestigious medical body is indicative of how

TABLA II.

Niños lactantes.

Especie segunda.	Alimentos y cantidades.	Horas. M. M. T. N
Recien nacidos de dos dias hasta ocho. A dos de los quatro que de esta edad puede tener una nutriz recien parida.	Tres mamadas al dia de leche fresca..................	.6...............2... 10
A los otros dos.	Lo mismo........	10.............6.. ...2
Sobrantes que toman leche.	Tres xícaras al dia de leche de burra recien ordeñada ó de cabras con tres partes de agua caliente azucarada...........	.6...............2.. .10
Que no la toman.	Tres tacitas de un caldo tenue ó de substancia de pan, arroz, &c. tenue y azucarada......................	.6...............2.. .10

Figure 3.4. Table of the food recommended to babies between two days and eight days old. In Santiago García, *Instituciones sobre la crianza física de los niños expósitos* (Madrid: Imprenta de Vega y Compañía, 1805). Biblioteca Nacional de España.

the society wished to present itself publicly. Although the Junta counted among its members learned women who were knowledgeable about infant feeding, it was certainly an outsider in the medical community. The diaries were intended to prove how rigorously the tests had been carried out in the foundling house under the governance of the Junta and how reliable the data they gathered was. At the time, the academy had great prestige, especially because of its secretary, Ignacio María Ruiz de Luzuriaga (1763–1822), who had studied in Paris and finished his dissertation in Edinburgh and was closely connected to other Spanish societies such as the Royal Society of the Basque Provinces of Friends of the Country.[53] The academy was becoming an authority in national public health. It reported on medical manuscripts and recommended or forbade their publication; was consulted by the government on epidemics, such as the ongoing outbreaks of yellow fever; recommended public health regulations; and intervened in polemics about health.[54]

In fact, the Junta had in the past successfully associated with other experts relevant to their work, in particular with the director of the Royal Laboratory of Chemistry, Pedro Gutiérrez Bueno, on two occasions.[55] One occasion was in relation to several tests carried out on textile dyes in the Montepío de Hilazas (see chapter 1), while the other was in relation to the trials carried out in the Madrid jails in 1790 to purify the dungeons' air. On that time, the trials were carried out not by the Junta but by the Señoras de las cárceles. Although the Señoras had been set up as the initiative of a priest—in order to alleviate the poor conditions of women in the Madrid jails—many of their members also belonged to the Junta, including the *curadora* of the foundling house, the Countess of Montijo. For instance, between 1788 and 1805, of the thirty-four members of the Señoras, ten belonged to the Junta.[56] During the mentioned trials, the Señoras intended to increase the content of oxygen in the dungeons of the Madrid prisons, and to that end they worked with Gutiérrez Bueno to take air samples and measure the quantity of oxygen with a eudiometer in the Royal Laboratory. On finding that the oxygen levels were very poor, they designed a trial aimed at finding the best method to "purify" the air. Over a period of twenty days, aromatic herbs (the traditional way of purifying the air) were burned in several dungeons, while in others vinegar was poured in different forms, either diluted with water or pure (which was framed as the "modern" way of disinfecting with acids). The trials were widely publicized in the official press (the *Gazeta de Madrid*) and other journals.

To be sure, the public association of the Señoras with the director of the Royal Laboratory benefited all the parties involved. For instance, it increased the visibility of Gutiérrez Bueno, who also worked for the government as an expert in public health and as a chemistry tutor of gentlemen, while demonstrating the benevolence of the Crown—which was a generous benefactor of the Señoras— to the people of Madrid. But it also undoubtedly highlighted the acquaintance of the female society with modern theories of disinfection and chemical practices. In the case of the foundling house and the baby food trials, I would suggest that the benefits also worked for both parties: the physicians who worked there benefited from the Junta, while the diaries of the foundlings' milk trials were used to cement the seriousness of the Junta in medical and savant circles.[57] In fact, the Junta mentioned precisely this rationale in their annual report of 1805, which was then forwarded to the Madrid Economic Society and the Crown: "The exact diary of these tests is the best testimony of how carefully it has been carried out, but the members of the Junta did not dare judge the results, so they passed the diaries to the Royal Academy, whose opinion we are still awaiting."[58] However, as the following shows, the Junta indeed dared to judge the results and acted independently of the academy's judgement.

On November 15, 1805 (after the diaries were sent to the Royal Academy of Medicine in June 1805, and while the annual report was issued), the Junta changed the diet of the goats involved in the experiment. Some would be fed by fodder of barley, stalk, and carob, while others would be fed with fresh grass.[59] In December, it was concluded that even with the new diet of fresh pasture, children were not faring well. The Junta then tested donkey's milk until May 1806. According to the Junta, this was proposed by one of its members, Doña Josefa María de Panes.[60]

In the meantime, the report of the Madrid Royal Academy of Medicine on the diaries of the trials arrived (on March 29, 1806).[61] Stressing gender differences, the academy clearly demarcated the territories of women and physicians. It did so in a way that revealed tensions between the two institutions. First, the academy apologized for the delay of more than a year but stated that this was apparently motivated by the refusal of the Junta to receive an inspection visit from the academy to the foundling house (which also suggests some possible tensions between the two collectives).[62] Then, the academy praised the way the Junta had governed the house, stressing their gender and their learned qualities: "[the Junta] has joined the sweetness of its sex with the lights of its class"; "each lady is a caring and tender mother," and each foundling "a cherished son."[63] Finally, in polite terms, it pointed out that the medical authority of the academy had been bypassed, as the Junta was testing donkey's milk (we do not know by which channels the academy learned of this): "But the Junta has relieved the academy of any obligation to do this work [by proposing that they themselves try out donkey's milk], and has administered donkey's milk, and a positive conclusion has met all expectations . . . aided by virtue, this ingenuity was as successful as art, and thus have the ladies of honor given further proof of their enlightenment and zeal and established an inviolable law in a matter of uncertainty."[64] The Royal Academy of Medicine adroitly detached itself from the Junta. Art was for the academy, virtue for the Junta. Without disparaging the tests, the academy assigned to the Junta wit, virtue, and zeal, but kept the art for itself.

However, in the long run, donkey's milk did not work either.[65] Three years later in 1816, the physician mentioned above, Anento, graphically expressed the dismay of the Junta and the doctors of the house: "I must confess that I am also amazed to see and have experienced the unfavorable effects that resulted from the experiments. Barely a child was preserved, neither suckling little or much, following an established method or not, directly from the goat's udder, or with a feeding bottle, diluted or pure."[66]

Notwithstanding, the Junta persisted in its search several years later, testing goat's milk from April to May 1812, and a tropical root, the *nuevo sagú* (*Amaranta arundinacea*), in 1817.[67]

Trials with Tropical Roots and Beyond

By 1817 the political situation of Spain had drastically changed from 1805. The intervening period had seen the Napoleonic Wars, the parenthesis of the liberal constitution in Cádiz, the outbreak of rebellion in the American colonies, and the reestablishment of the absolutist regime by Fernando VII in 1812. The sources nevertheless suggest that the Junta kept up its subterfuge of actively supporting research in food substitutes while concealing its agency. There are two moments I would like to discuss before delving into the trials the Junta carried out with tropical roots. One is the report delivered in 1816 by the then-director of the Royal Academy of Medicine, Ignacio María Jáuregui, and the other is a draft of a letter to the Madrid Economic Society. Both support the thesis of the independence of the Junta from medical doctors, and both suggest possible strategies used by the Junta to delimitate their responsibilities.

The report by Ignacio María Jáuregui (1816) was prompted by an inspection visit of a commission of the Royal Academy of Medicine to the foundling house the year before, probably to check the wards and the state of the children and wet nurses. It can also be traced, however, to the Junta's wish to again consult about the trials with goat's milk.[68] Jáuregui praised the Junta's governance of foundling house, but interestingly, his report elaborated at length about the properties of milk as the only substance adequate for infant diets. As Jáuregui wrote, milk was the only substance that met the "principle of animalization" and offered less resistance to taking "the form and qualities of human life." He concluded that any other substitute, including tropical plants, was to be ruled out.[69]

A year later, however, the Marquise of Casa-Flores brought from Cuba a tropical plant, the *nuevo sagú* or arrowroot, along with a recipe for preparing a type of flour with its roots. Apparently, a pap with this flour, water, and sugar was used in the foundling house of Havana to feed the newborns.[70] Despite the opinion of Jáuregui, the Junta decided to carry out a trial with some of the children. Moreover, before obtaining any results, the Junta wrote to the Madrid Economic Society to explain that the trials with the *nuevo sagú* had begun with good prospects, proposing the acclimatization of the plant into Spanish soil.[71] A draft of this letter suggests how the Junta carefully crafted its image. One sentence—"the trial [*experimento*] has begun under the direction of the countess"—was crossed out and replaced by a note that "the results will be given by the house doctors."[72] Although we do not know who edited the draft, the change suggests that the Junta wanted to remain behind the scenes, perhaps letting doctors have the role of protagonist, or possibly as a kind of insurance: if the trials didn't work out, the doctors could be held responsible.[73] The documents don't allow us to rule out either of those possibilities. However, whatever the reason was, the result is that the Junta placed themselves under the technical direction of the doctors, effacing their own agency.

However, in spite of the great expectations, the *nuevo sagú* would not be the solution to the hunger among the children in the foundling house. The trials were conducted in September 1817, with twenty-one sick babies and a healthy newborn. The house physician at the time, Pascual Mora, wrote a weekly letter to the Junta explaining the development of the children. Some letters are laconic: "nothing can be said," the doctor wrote, but in others, he described characteristics of the flour, such as "it is similar to starch" and "it has a sweet substance and gluten."[74] The final letter, sent on October 3, 1817, included a table in which Mora specified how many of the children died every day. Within eighteenth days, all perished, except the healthy one who recovered after he was given to a wet nurse. Mora offered a physiological explanation for the failure in terms of the weakness of infants' stomachs and the insufficiency of their "lactic vessels." He then recommended trying a combination of the *nuevo sagú* with goat's milk, because the latter produced "diarrhea and depurative eruptions" while the former stopped them.[75] This letter suggests that he needed the permission of the Junta to try any new tests, which the Junta gave, as a week later, a new trial combining goat's milk and *nuevo sagú* flour was carried out with the infants then present in the infirmary. Again, the results were catastrophic.[76]

The results on the trials made it into medical treatises on infant growth. Three years later, in 1820, Tomás Mora presented the account of the trials to the Royal Academy of Medicine, with a chemical analysis of the arrowroot and a discussion of the chemical composition of other tropical plants.[77] In 1828 he published a massive three-volume work on the development of children, in which he included a special supplement on the benefits of milk, the trials with the *nuevo sagú*, and discussions of other tropical plant starches used for food in indigenous cultures.[78] However, in Mora's telling, women's role was limited to preparing the flour.[79]

I would like to end this chapter with a late testimony that further demonstrates the Junta's agency in the trials. In the summer of 1828, 150 babies crowded in the wards of the foundling house, with each wet nurse having to suckle three babies.[80] The women then responsible, the Countess of Alagón and the Countess of Casa-Sarrià, considered restarting the trials with goat's milk, as they feared that all the children would perish. However, they decided to consult first with the former president of the Junta, the Duchess of Benavente.[81] Having lost her voice for some time, the Countess of Alagón penned a short letter instead of visiting the duchess. Her aphonia was truly fortunate for us, as now we know that the president of the Junta registered the diary outcomes of the trials: "And as you had the curiosity to take notes of all that was done, and on the novelties that occurred with the children. . . . We would be thankful if you could please us lend your notes, which we will return immediately, as they will be very useful for the new trials."[82] The aging Duchess of Benavente, then in her eighties, could no longer find her notes, but she did remember that all the children passed away. In her words: "In spite of the ladies' care . . . all the hopes to achieve the happy results of other foreign countries faded."[83] The Countess of Alagón tried then to find more about the past trials elsewhere in the house. As she wrote to her companion, she did so in order "that the doctor and we will know."[84] Finally, due to the risks, the countesses decided not to carry the trials out, despite the fact that the doctor had already picked the children. As late as 1828, women in this case still seemed to have the last word.

In Spite of the Ladies' Care

Although patchy, the story of the trials suggests that the Junta hoped to resolve a pressing issue that occupied people from all walks of life, namely, possible alternatives to breastfeeding. What made women's milk so particularly suitable for newborn babies was the underlying question, which, of course, could not be answered until much later. Yet, the Junta felt well-prepared to embark on trying different solutions. They were aware of the literature on infant food, they researched what had been written about it in their commissions on physical education, and members abroad brought news on methods that seemed to work. The Junta also felt that their efforts were legitimate, as other societies and physicians were also testing alternatives in foundling houses. Just a year after they took over the foundling house, the Junta began funding, witnessing the trials, and registering and discussing the results among themselves and the doctors of the house.

Still, the arena the Junta was stepping in was disputed by other male experts, in particular by the Royal Academy of Medicine. The collaboration was thus complex. On the one hand, the approval of one of the most important medical authorities gave public authority to the Junta. On the other hand, based on the information available, it seems that both societies allocated different spheres for themselves. The knowledge and skills of the Junta were gendered as feminine, in the sphere of care and motherhood; the knowledge and skills of the Royal Academy of Medicine as masculine, in the sphere of "proper" medicine.

One question remains, though. Were the trials the results of the stubborn ambition of the Junta rather than a possible path to saving children from a sure death? Were the women using the babies simply to test medical novelties? Such an interpretation would be at odds with the Junta's stance in the notorious discussion about the use of goats in vaccine trials. In March of 1803, Juan José Heydeck, professor in the Reales Estudios de San Isidro of Madrid, claimed that goats could be used instead of cows to obtain the serum for inoculation against smallpox. Given the lack of volunteers to try the goat's serum, the Protomedicato (the maximum health regulation authority) ordered that it be tested on the foundlings. The Junta, however, refused to obey the order. According to the Junta, no favorable trials had yet been carried out that would justify this risk to the infants: this would be a "trial that, because of the harms and inconveniences that it might cause to the children, at present has no favorable probability."[85] This refusal suggests that the Junta carefully balanced the medical evidence of successful cases with the risks a treatment might pose. Experimenting was dangerous to the children, but finding a substitute for breastfeeding would have been a triumph given the desperate situation of the house.

We leave the busy Junta now to explore in the next two chapters other examples of alliances between reformers and women improvers: first in the education of children, and then in rural economy.

4

Learned Daughters

In 1787, just as the Duchess of Benavente was busy with the Junta de damas, she commissioned from Goya a portrait of her growing family.[1] Goya depicted the four children, aged between two and five years old, as if he just surprised them playing with their parents (figure 4.1).[2] The duke is portrayed handing a spinning wheel to the eldest, Josefa, with one hand while protectively setting the other on the armchair, where his wife sits. The duchess, carefully dressed in an English buttoned attire, holds a small book in her lap and embraces little Joaquina, who clasps a fan in one hand and a toy in the other. At their feet, the youngest, Pedro, drags a tiny carriage, while the heir, Francisco de Borja, pretends his father's baton is a horse, perhaps a reminder to the spectator that he stood to inherit his father's regiment but also a sign of how the duke set aside the symbol of his power to play with his children.[3] The atmosphere of cheerfulness is enhanced in subtle ways—the half smile of the duchess, the puppies at the feet of Josefa. Goya represents here an ideal enlightened family, in which spouses are united by a companionable marriage and educating the offspring is a pleasant and rewarding task.

In the Enlightened imaginary, one of the most important and "sweetest" maternal duties was to procure a good education for their daughters.

Figure 4.1. Francisco de Goya, *The Duke and Duchess of Osuna and Their Children*, 1787–1788. Oil on canvas, 225 x 174 cm. Madrid, Museo Nacional del Prado. The Duke of Osuna, Pedro Téllez-Girón (1755–1807); his wife, the Countess-Duchess of Benavente María Josefa Alonso de Pimentel y Girón (1752–1834), and their four children: Francisco de Borja (1785–1820), the subsequent Duke of Osuna; Pedro de Alcántara (1786–1851), María Josefa (1783–1817), and Joaquina (1784–1851), Museo del Prado. © Archivo Fotográfico Museo Nacional del Prado.

However, what this meant was a matter of protracted controversy. During the ancien régime, education had a utilitarian character, meaning that each estate was to be educated according to their future social role. The question of women's education thus implicated deciding what these roles were, what women should learn in order to fulfil them, and how—all of which were issues that raised never-ending debates.[4] The education of high-ranking girls was expected to include at least religion, history, geography, domestic economy, and social accomplishments (music, embroidering and other needlework, painting, and dancing). A knowledge of French and Italian, as well as a shallow knowledge of the sciences, was also advised.[5] Poor girls, on the other hand, if they were educated at all, were taught religion, textile work, and sometimes, the rudiments of reading, writing, and arithmetic in schools funded by patriotic societies. In the Hispanic colonies, the situation was similar—well-off girls were often sent to private houses (houses of Amigas o Migas, that is, "friends"), where they learned reading and writing, Christian doctrine, and the tasks of domestic labor. Literacy rates were also very low.[6] As the century progressed, education came to be understood in more egalitarian terms, but nonetheless, up until even the late eighteenth century there was a great chasm in Spain between the massive illiterate rural population and the privileged urban one.[7] Only around 13 percent of women in Spain could read, compared with 27 percent of women in France and approximately 50 percent of men in Spain.[8]

In recent years, the picture of the allegedly poor, superficial education of high-ranking women has been nuanced and questioned.[9] The education of the elite transcended the limits of the personal, deeply involving the dynasty's prestige and the supposed leading roles of noblewomen in society. Not only did aristocratic women often hold paid positions at court—they were frequently expected to be socially and politically influential, as many of them would marry ambassadors, diplomats, and other high ranked officials, or businessmen. They were also expected to advance the family's power, to establish alliances and secure pensions for their kin, and to be models of behavior.[10] Receiving a broad education was key to their success. Furthermore, the cases of exceptional women—such as Cayetana Gúzmán de la Cerda, Rosario Cepeda, and Pascuala Caro y Sureda (discussed in chapter 1)—illuminate how learned daughters served multiple purposes, from fulfilling the political goals of ambitious fathers to supporting patriotic claims about how modern a country was.[11]

The ways in which these women achieved their outstanding education depended on familial circumstances. They might have shared their preceptors with their brothers, as was the case with the girls of the Osuna family and of Josefa Amar (whom we encountered in chapter 1), or they might have been encouraged by the male kin of the family, as attested by the case of the Canary islander María de Betancourt, sister of the famous engineer Agustín de Betancourt.[12] But along with more formal lessons, there were other, informal ways of learning. This chapter explores the role that the new juvenile literature and pedagogical toys might have had in the education of Spanish girls in sciences, a poorly charted territory to date.

As the century progressed, children's books would occupy pride of place in elite nursery rooms. To be sure, books addressed to children had also been available in previous centuries, for instance, *The Adventures of Telemachus, Son of Ulysses* (1699), or Comenius's *Orbis Pictus* (1658). Chapbooks, chevalier novels, romances, and particularly fables were popular reading among the youth.[13] However, books marketed in ways to please children and their parents were a genuine late eighteenth-century innovation.[14] They were usually of small size (often in octodecimos and sextodecimo), with illustrations, wide margins, and spaced lines. Yet their most salient feature was that they pretended to be instructive and entertaining at the same time. The books' characters, structure, and language were carefully chosen for "instructing with delight." Based on the idea that a child could learn almost anything, without effort, if it were packaged in the right way, the new juvenile literature was aimed at both boys and girls.[15] The idea of course was not new—it had been articulated by Locke in *Some Thoughts Concerning Education* in 1693, but now publishers fully recognized its market potential.[16] In particular, contemporary science was a favorite topic in such books. Tales, poems, and short plays easily incorporated lessons from natural history, botany, geography, chemistry, mineralogy, and agriculture. One could find an astronomical discussion while reading the adventures of two shipwrecked children in the Pacific Ocean and accurate botanical descriptions of the landscape in a tender tale about a mother and her infant.[17] In the French *Les Voyages de Rolando et de ses compagnons de fortune autour du monde* (The travels of Rolando and his companions around the world) by Louis-François Jauffret, children learned geography, history, and natural history as Rolando and his companions traveled around the world.[18] In the British *Children's Magazine* (1799), a tiny and beautifully engraved journal, children could read about natural history—the lives of insects, mammals, birds, and fish—and geography, as well as stories of how new commodities such as tea, coffee, cotton, and sugar arrived from the English colonies.[19] *The Young Gentleman's and Lady's Magazine* published articles on botany and natural history, while the more serious *Young Gentleman's Magazine* contained geography, history, and moral precepts extracted from Jean-Jacques Fillassier's *Eraste ou l'ami de la jeunesse* (Erastus, or the friend of youth) (1773).[20]

Books fully devoted to teaching sciences were also popular. The best-known among historians of science thanks to the groundbreaking article by Jim Secord is probably the tiny *Philosophy of Tops and Balls or The Newtonian System of Philosophy Adapted to the Capacities of Young Gentlemen and Ladies* (1761), but hundreds of other publications for children placed scientific knowledge in familiar settings.[21] As the frontispiece of the *Juvenile Magazine* (1788) showed, girls and boys would be guided to the temple of the sciences just as joyfully as if they were playing. Indeed, the wealth of pedagogical material culture was well reflected in these books. The little characters played cards (natural history cards, of course!), lotteries (grammatical lotteries), table games (for learning morals), and puzzles (for learning geography). They had many hobbies: they collected shells and minerals, experimented with gases, and made cardboard planetary models.[22] They also owned globes and instruments to understand physical phenomena and natural laws. Children's books thus also fostered the use of toys and artefacts to stimulate children's curiosity and help them learn in arguably effective and lasting ways.[23]

Furthermore, scientific education supposedly reported other benefits. Notice for instance some of the virtues ascribed to the study of botany: "Botany cultivates a taste in young persons for the study of nature, which is the most familiar means of introducing suitable ideas of the attributes of the Divine Being, by giving them examples of the order and harmony of the visible creation. . . . Botany contributes to the health of body and cheerfulness of disposition by presenting an inducement to take air and exercise."[24] The mutual reinforcement of moral, intellectual and body instruction has been brilliantly discussed elsewhere.[25] Here, I would like to stress another point: that grown-ups were also included in the formative package. Admittedly, all children's literature dialogues in one way or another with adults. Children's books are certainly a meeting ground for readers of different ages and even bring people face-to-face with their younger self, "cross-writing the child and the adult," as Mitzi Myers put it.[26] Yet, in the eighteenth-century this intergenerational aspect is mostly devoted to showing parents and tutors the proper ways of educating: how to instil in children the taste for learning, how to drive an instructive conversation, and how to look at nature to bring home moral lessons.[27] Children's books also showcased models of adult behavior within and outside the family: how the relationship between spouses should be structured, which values should be respected, and what the different masculine and feminine roles should be.

The belief in the power of juvenile literature to effect social change was particularly exploited by French revolutionaries, who used the genre to develop the new citizen—free of religious prejudices, steeped in the correct moral values, and with knowledge of the empirical sciences.[28] The possibilities of the juvenile genre also did not escape Spanish reformers, who devoted time and efforts to write, translate, and adapt children's tales to their political agendas. This chapter therefore delves into the first juvenile journal published in Spain (in fact, the only one until the 1830s), the *Gazeta de los niños o Principios generales de moral, ciencias y artes, acomodados a la inteligencia de la primera edad* (Children's gazette or general principles of morality, sciences, and arts adapted to the intelligence of the first years, 1798–1800).[29] Inspired by a French journal, the *Gazeta* was authored by two influential civil servants, the Canga Argüelles brothers. We will see how they envisioned a broad scientific education for girls in order to transform a supposedly backward Spain. We will then contrast the Canga Argüelles's writings with some Spanish female authors, who pictured active mothers and governesses keenly involved in children's education, which in turn inspired and were inspired by high-ranked women. We will end with an episode from the life of Duchess of Osuna, ten years after the Goya's portrait discussed above, which gives some clues regarding the broad education she provided Josefa and Joaquina, with the aim of preparing them to play an active future role in society and politics.

A Journal for the New Spanish Citizens

The pioneering study of Paula Demerson shows that more than six hundred titles were announced in the main Spanish journals between 1740 and 1808.[30] But in contrast to Britain, in which a generation of women—such as Anna Barbault, Sara Trimmer, Maria Edgeworth, Charlotte Smith, Priscilla Wakefield, Elizabeth Helme, Jane Marcet, Eleanor Fenn, Mrs. Pilkington, and

Mrs. Sherwood to quote just a few—wrote extensively for juvenile audiences, in Spain as in France most authors were men with political and social aspirations.[31]

Certainly, the fact that children's male authors greatly outnumbered female ones reflected the structure of the Spanish book market.[32] But it also testifies to how this genre enhanced men's public careers and writers' claims of social utility. For instance, the translator of *The New Robinson*, a well-known playwright, Tomás de Iriarte, belonged to an influential family of diplomats and was close to the aristocratic circles of the Marquise of Castelar, the Duke of Villahermosa, and the Duke of Osuna.[33] Valentín de Foronda, an active fellow of the Royal Society of the Basque Provinces of Friends of the Country who wrote an instructive chemical dialogue for children, would become general consul in Philadelphia in 1804.[34] Many other juvenile writers earned their living in aristocratic houses. For instance, the secretary of the Duke of Osuna, Manuel María de Ascargorta, translated a book referred to as "Buffon for Children," while José Viera y Clavijo, tutor of the children of the Marquis of Santa Cruz, translated "The Loves of the Plants," the renowned botanical poem by Erasmus Darwin about Linnaeus's plant classification system. He also composed several didactic poems about newly discovered gases and a little astronomy manual for children.[35] Neither should we underestimate the fact that juvenile literature was less risky than other genres. In the context of strong state censorship and a still-active Inquisition, it was safer to devote one's pen to pedagogical literature, in which there was some room to transmit one's own ideas to young generations, than writing political essays.[36]

In fact, the *Gazeta de los niños* was inspired by the post-revolutionary French journal *Le Courier des Enfants* (1796–1799), authored by Louis Jauffret (1770–1840), an emblematic figure in Parisian intellectual elites and one of the first to earn a living through scientific popularizing.[37] Jauffret had tight connections with the scientific elite, such as Georges Cuvier and Antoine-Laurent de Jusseau, who helped him with his juvenile books on anatomy; he also created the first "anthropological" society, organized popular botanical promenades, and designed pedagogical games such as the Jeu zoologique et géographique destine à l'amusement et á l'instruction de la jeunesse (Zoological and geographical game intended for the amusement and education of young people, 1798–1799)—fifty-two animal cards that children had to match with the correct nation on a map of the world.[38]

José (1771–1842) and Bernabé (1778–1812) Canga Argüelles fully embraced the pedagogical values of the *Le Courier des Enfants*, in particular the effort to appeal both to reason and emotion to drive home morality lessons rather than resorting to religious arguments. The brothers were well-known in reformist circles and among the scientific elites. They worked in the Treasury Department (Departamento del Fomento General del Reino y de la Balanza de Comercio) and were deeply involved in the politics of the time.[39] They debuted in the literary world as Hellenists, later moving to political economy and writing commentaries to Pedro Rodríguez de Campomanes's influential *Discurso sobre la educación popular* (Discourse on popular education) and on the census of the population.[40] In 1798 José was appointed as a civil servant of high rank in the Treasury Department, where he would developed a fruitful career and come to be considered today a key Spanish economic thinker.[41] He would join the liberal party during the hectic period that followed the Napoleonic invasion; with the ascension of Fernando VII he would go into exile in England, where he would write more than three hundred essays, a treatise on tax revenue, and a highly praised dictionary on the same topic.[42] The younger brother, Bernabé, was also interested in the sciences—in particular, in geology and mineralogy.[43]

The importance that a juvenile journal was given in reformist circles is further evidenced by their choice of the publishing house. The *Gazeta* would be published by the innovative presses of the Sancha family, which was probably keen on the idea of reaching a broader public than with books. Over several generations, the Sanchas had strengthened ties with the French book business and were highly committed to the sciences, publishing, for instance, Fourcroy's journal on chemistry, Lavoisier's treatise on the new chemical nomenclature, and the *Encyclopédie Méthodique*.[44]

The necessity of fomenting national technical knowledge was a long-standing concern of the Canga Argüelles brothers. Consider for instance their project of a museum of Spanish products, the Depósito Industrial (Industrial warehouse). The museum would classify all the goods therein with a "precise system" that would serve to direct future policies. It would include products from rural industry (wool, silk, wax, sugar, woods, vegetable, and tinctures), urban industry (things made by artisans, such as marble makers, potters, and carpenters), and materials produced by "chemical and mathematical-physical arts."[45] According to the brothers, a review of the collections would show both "the grandiose picture of the industry of Spain and its colonies" and "the empty spaces that a wise management could fill in a short time."[46]

As a first step toward filling the empty spaces, the Canga Argüelles set about publishing the *Gazeta*. According to the foreword, they were confident in the power of education: "Ancient history offers famous examples of the power of education that changed the character of whole nations under the gaze of the same philosophers who reformed them."[47] The goal of the Canga-Argüelles was ambitious. In only the first two issues, they promised to teach "some general ideas of natural history, some principles of geography and geometry, the farming and spinning of hemp, a treatise on the influence of education in children's physiognomy, the theory of heat and theory of the gases that formed our atmosphere, some preliminary notions of the natural history of birds, and several maxims and moral tales."

For their reformist adventure, the Canga Argüelles needed both boys and girls. Moreover, although the action in the *Gazeta* advanced through dialogues between two fathers and children of both sexes, there were particular instances in which girls were addressed directly, suggesting that the authors wanted to emphasize that girls should also be educated in empirical sciences.

Enticing Girls to Learn Empirical Sciences

The structure of the *Gazeta* was the one common in juvenile journals of the time: a miscellanea of short stories, plays, riddles, and poems framed in a familiar conversation—in this case, between two families who spent the summers at their country houses in Toledo and the winters in Madrid. The two scenes, the rural and the urban, allowed the authors to switch easily between topics in the natural sciences and political economy.

All conversations in the journal are directed by the two fathers, who take advantage of every little occurrence to instruct their children. As the authors explain in the prologue, "in the streets, in the ambles, everywhere, they [the fathers] find the means to increase the knowledge of their children with experiences and lessons."[48] In the summer, they chat about natural history and agriculture; nights spent beneath the stars on the riverbank are dedicated to astronomy, meteorology, and entertaining the children with "moral stories." In the urban winters, they discuss commerce and industry and visit chemical laboratories and friends' botanical gardens. Mondays are devoted to the Natural History Museum, where they delight in its stuffed animals, and the impressive collections of minerals and insects. They pause at each display case to observe all the details and read some pages of Buffon's *Histoire naturelle*, which had just been translated into Spanish, in order to note the profitable uses of everything.[49] At home, the children and fathers experiment with domestic things, from tea pots and soap to study gases, to glass prisms for diffracting white light.[50] Yet, above all, the Canga-Argüelles wished to convey the lesson that children must learn to cherish the arts: "The simplest work of the arts demands from the professor so much knowledge . . . that only a blind custom can make one to look at them with indifference."[51]

We cannot rule out that the literary fathers represented the alter ego of the Canga Argüelles (as both had quite remarkable personalities), but they were also following a tradition typical of pedagogical books, in which a male tutor accompanies the child day and night. They also reflected a social reality, as elite children likely had preceptors who lived with the family. From the mid-eighteenth century on, the most famous literary male tutor is found in Rousseau's *Emile, ou de l'education* (*Emile, or On Education*) (1762), but there were many other examples, such as Jean-François Dubroca's *Entretiens d'un père avec ses enfants, sur l'histoire naturelle* (Conversations of a father with his children about natural history), in which the leading educative role was played by a learned man.[52] However, the great difference in comparison to *Émile* is that in the *Gazeta* girls are educated along with boys, in the same topics and in the same experimental and rational manner, in stark contrast to the education Rousseau envisioned for Sophie, the to-be companion of Emile.

From the very beginning, the authors explained that it was a girl who inspired them to write the *Gazeta*: observing that a girl found great joy in reading Berquin's journal, they decided to publish a Spanish one. The fictional girls are educated together with their brothers, and when the girls miss a lesson, the boys teach it to them later. Tellingly, the few times that this occurred was when the elder boy visited sites of empirical research with his father, places in which women were not often seen, such as Madrid's chemical laboratories. That the Canga Argüelles brothers understood that girls needed, despite this, to be taught in chemistry is apparent in the changes that they made to a well-known tale by the already mentioned Jauffret.[53]

In *Voyage au Jardin des Plantes*, Jauffret explained the galleries of natural history, the zoo, the botanical garden, and the lecture theater of the Parisian Muséum National d'Histoire Naturelle through conversations between a boy called Gustave and his tutor. The book is organized around several journeys. On the second day, Gustave observes a crowd of ladies at the doors of the lecture theater, which was used to hold public courses in chemistry that were very well attended by elite Parisian women.[54] Surprised, Gustave asked his tutor: "What are these ladies doing? Perhaps they want to learn chemistry?"[55] The tutor's answer makes clear that the ladies indeed came to learn the new chemistry. In three sentences, Jauffret summarizes its most salient features —its novel nomenclature, the idea that substances are composed of elements, and the discovery of the chemistry of gases: "The ladies came to learn that the niter or saltpeter is today named potassium nitrate, that cooking salt is called muriate of soda; they come to learn to decompose the elements; to change water into air and air into water; to know the two bases of common air, which are oxygen and nitrogen gas."[56] Jauffret then describes some of the classical experiments that the well-known chemist Fourcroy used to carry out in the amphitheater. The original chemistry lesson in the *Voyage* lasts a page and a half.

The Canga Argüelles brothers changed the tale to serve their pedagogical goals by situating the action in Madrid, introducing a girl, and extending the chemistry lessons over three chapters. The new setting not only promoted Spanish institutions and experts but also reinforced the overall message. Other issues of the *Gazeta* accurately describe Spanish institutions, settings, and contemporaneous events, for instance, the wards of the Natural Science Museum, the Plaza Mayor of Madrid, or the visit of an itinerant zoo. By setting the action in real scenarios and spotlighting events that were known to readers, the whole educative project took on an aura of plausibility. In Madrid there was no equivalent to the Parisian lecture theater, nor were women well received in scientific institutions except on special occasions.[57] So in the *Gazeta*, one of the children of the family, the boy Augusto, happens to be in Turco Street in Madrid with his father at the very moment when the chemistry lessons given by "Chabenau" in the glass factory are about to begin.[58] The boy enjoys the lecture so much that he decides to explain it in correspondence to his cousin Luisita. Readers of the *Gazeta* may well have known the spot, which was in fact very near to the old laboratory of François Chabenau (1754–1842), a celebrated French professor who had previously lectured in Madrid.[59]

In the same manner, introducing a female character was thought to be an effective way of enticing girls to learn chemistry and perhaps encouraging their tutors to teach them. As the authors stated in the foreword mentioned above, "Children receive the instruction with more pleasure when it is communicated by other children."[60] This was the same rationale that underpinned the popular genre of "science for women," in which a male tutor instructed a female pupil (chapter 1).[61] The dynamics were the same between the male tutor and the female student—Luisita, as all female pupils, was delighted with the classes, and Gustavo, as all male tutors, was enamored of his pupil's intelligence, perpetuating the dynamics of male expert versus female lay practitioner but also reinforcing the idea that women were entitled to learn science.[62]

What, then, was the chemistry that boys and girls should know? It was the one considered useful: the properties of the gases and their importance in breathing and combustion and the alkalis and their uses in bleaches and soaps. In his first letter to Luisita, Augusto describes heat (*calórico*) and why different bodies melt at different temperatures.[63] He explains how the air we breathe is in fact composed of various gases and describes the well-known experiment of the glass-covered torches that burn until they consume all the available oxygen. Augusto finishes with the hope that someday Luisita will learn how to prepare several gases and experiment with them. The following month, he reviews with her hydrogen and its uses in balloons.[64] He recommends that Luisita procure some gas samples and details how to make domestic "bombs" with hydrogen. And in the next issue, the chemistry lessons continue in a different format. Augusto and his father suddenly find a man lying in the street surrounded by curious people seemingly incapable of helping him.[65] The father applies a small flask containing a volatile substance to the man's nose and the man immediately comes to life. This serves to introduce the "*alkalis*" (highlighted with italics in the original), particularly "ammonia, potash, and soda," which had "innumerable uses." The father explains the utility of potash in the formation of bleaches and how laundresses use it to remove stains. The lesson finishes with Augusto promising to write Luisita about all that he learned. Useful chemistry was found in other parts of the *Gazeta* too, for example when the journal discussed the chemical compositions of soils and the effect of fertilizers.[66]

Chemistry was not the only scientific knowledge that the Canga Argüelles brothers wished to impart to Spanish children. In the *Gazeta*, one finds lessons on botany, natural history, physics, and agriculture with a stress on their practical applications. In the tale called "The Garden," also inspired by Jauffret, Amadeo and his father promenade amid the garden of a friend and chat about mulberries, bananas, aloes, and cotton.[67] The fictional children learned (apparently with fascination) the names of all the pieces in the Piamonte loom, the plough, and the cotton mill, and of course, the hemp machine invented by the Catalans Francesc Santponç i Roca and Francesc Salvà i Campillo, used to separate the soft and hard parts of hemp.[68] Children were informed about how castor skins were made into hats and cocoons into silk, as well as about the supposed advantages of the free market.[69] As the astonished children wander through the marketplace, full of toy shops, instrument makers, and peddlers, the father exclaims, "All these, my friends, must make you love the society and the union of men, respect the arts, agriculture, and commerce [. . .] and work with all your strength to be useful in something to those who toiled so much to be useful to you."[70]

To what extend the *Gazeta* succeeded as a commercial venture is not yet clear. For instance, although in the first issue the authors expressed their intention to include pictures if they have enough subscriptions, they never did so, perhaps for lack of funds. Also, the enterprise only lasted two years. One possible reason is that the younger brother, who seemed to carry most of the work, fell ill; however, it was not until later in the century that a youth magazine was relaunched, which speaks perhaps against the economic viability of such a project in Spain.

Yet, the *Gazeta*, in spite of its French revolutionary roots and its aims to reach a big audience, reflected a common situation in high-class households: a particular relationship between preceptors and their children. Moreover, in the *Gazeta*, mothers don't give lessons, but they do facilitate children's instruction.[71] There is an interesting passage in the *Gazeta* in which the mothers contract a showman with a magic lantern to come to their home and project images of the solar system onto the walls of the parlor for the children to learn. We will consider these aspects in more detail in another episode involving the merry children we met at the beginning of this chapter, namely, the four Osuna siblings ten years later.

Preceptors and Elite Mothers

April 20, 1797, was a busy morning at the palace of the Duchess of Osuna.[72] The children portraited by Goya were now four teenagers, expected to sit for a public exam in the house. The same care that the duchess put into publicly presenting her family was deployed into the education of the children and the selection of their tutor. Almost ten years had elapsed since the duchess had hired Diego Clemencín as her children's preceptor. The duchess chose him after a thorough search among her friends, who highlighted Clemencín's vast knowledge in humanities and sciences and his good-natured character.[73] To be sure, the children would have other masters and mistresses, including the well-known painter Agustí Esteve for drawing, and the governess Mme de Saint-Hilarie for the girls to learn French, needlework, and music.[74] But Clemencín would unfailingly accompany the children through all their formative years. Flattering the duchess in his acceptance letter, he wrote, "There could not be anything more advantageous than serving a Lady of such enlightenment, who has such a splendid way of thinking about the education of her children."[75]

The morning of the exam was planned to the minute. A leaflet was printed with the names of the children, beginning with the girls: *Public Exam of the Mistresses Doña Josefa y Doña Joaquina Girón y Pimentel and the Masters Francisco y D. Pedro, Their Brothers.*[76] Clemencín would open the session with a short discourse in which he explained to the numerous spectators his pedagogical method as well as what could be expected of the children.[77] These latter would first answer questions from the audience in arithmetic and history, with Josefa, now aged fourteen, also answering in geometry. To show how much grammar they knew, they would play a "kind of lottery." Then, natural history and physics would have a turn. In a room devoted to natural history, the boys would explain the "most notable parts of the three kingdoms," while in a space devoted to physics and scientific instruments, they would experiment with a pneumatic machine (a device that created a vacuum and served to prove the properties of gases, among other things) and an electric machine (a device that charged objects with great amount of static electricity).[78] The new paintings that the duchess commissioned from Esteve the following year reflected the children's scrupulous education: Joaquina was portrayed with a terrestrial globe, Josefa with an embroidering table, Francisco de Borja with a telescope, and Pedro pointing to a natural cabinet that sheltered marine fossils, brittle starfish, seashells, and corals (figures 4.2, 4.3, and 4.4).[79]

Performances and portraits displayed the social status of the family and showcased the values that it championed. The Osuna family's public display

Figure 4.2. Agustín Esteve, *Joaquina Téllez-Girón, Future Marquise of Santa Cruz*, 1798. Oil on canvas, 190 x 116 cm. Madrid, Museo Nacional del Prado. © Archivo Fotográfico Museo Nacional del Prado.

Figure 4.3. Agustín Esteve, *Francisco de Borja Téllez-Girón, the Future Tenth Duke of Osuna*, 1798. Oil on canvas, 191 x 117 cm. Colección Masaveu, CM-506. Image from Virginia Albarrán, *El desafío del blanco: Goya y Esteve, retratistas de la casa de Osuna. A propósito de la donación Alzaga* (Madrid: Museo Nacional del Prado, 2017), fig. 20, page 27.

Figure 4.4. Agustín Esteve, *Josefa Manuela Téllez-Girón, Future Marquise of Camarasa*, 1798. Colección Fundación Casa Ducal de Medinaceli, Casa de Pilatos. Image from Virginia Albarrán, *El desafío del blanco: Goya y Esteve, retratistas de la casa de Osuna. A propósito de la donación Alzaga* (Madrid: Museo Nacional del Prado, 2017), fig. 14, page 22.

of natural collections, scientific instruments, and pedagogical toys vividly manifested its wealth, worldwide networks, and acquaintance with the new empirical sciences.[80] Yet in the hands of the children, the objects also spoke volumes about the duke and duchess's endorsement of new pedagogical methods of "learning by way of amusement." They also demonstrated their support for a broad education for girls, and very particularly, their endorsement of the education of their daughters in topics not traditionally in the female curriculum, such as geometry, grammar, and the empirical sciences.

These public spectacles also especially enhanced the image of the duchess as a "lady of enlightenment," as Clemencín put it, who so meticulously directed the education of her offspring. Further proof of the duchess's interest in the education of her children is found in an eighty-eight-page manuscript held in the Madrid National Library. It is titled *Project de Bibliothèque dressé d'après les notes remises par S.E. Madame la Duchesse d'Osuna* (A library project prepared from the notes sent by the Duchess of Osuna) and lists more than sixty-four hundred titles.[81] An unknown correspondent of the duchess in Paris classified the books under several epigraphs, specified their prices, and added useful notes for purchasing them. The great majority were novels, but 770 dealt with "morals and philosophy." We find books on the genre of "science for women," such as Fontenelle's *Entretiens sur la pluralité des mondes*, the French translation of Algarotti's *Il Newtonianismo per le dame*, and Maupertuis's *Venus physique*, in addition to classics such as the complete works of Locke, Condillac, Nollet, Pluche, and most of the fashionable children's educational books addressed to both girls and boys, especially those written by female authors.[82]

To be sure, the titles show the interest of the duchess in the genre; they also highlight the fact that children's books didn't need to be translated to be purchased. In fact, French books circulated broadly, bought either in Spanish bookshops or directly in Paris. We know, for instance, that the Duchess of Osuna wrote to the aforementioned publisher Sancha asking for books authored by Berquin and other children's authors that he sold in his printing house.[83] Also, the *Gazeta* wrote monthly reviews of recent pedagogical publications and included the addresses in Paris where they could be bought. For example, the journal mentioned the French translation of the British author Sara Trimmer's *Easy Introduction to the Knowledge of Nature*, in which a mother teaches two little children natural history, as well as a French geography manual, in which a mother teaches her son geography and history and the use of globes and maps.[84]

In other words, elite families had access to the wealth of French juvenile literature as well as to comparable British literature through French translations. Nonetheless, the work of some French female authors, such as Mme de Genlis, Mme d'Epinay, Mme Campan, and Mme Leprince de Beaumon, were translated into Spanish with great success.[85] The next section shows how, in books authored by women, the principal character is a governess or a mother, which poses interesting twists to the question of female agency in the education of children.

Female Children's Authors

Scholars have stressed the differences between female and male children's writers. Accordingly, female authored books were crucial in construing ideas about an "enlightened form of motherhood," giving mothers a more active role in children's education. Women authors often defended a broad curriculum for girls and reading lists more extensive than merely classical educational books, in accordance with their ideas that women themselves benefited from a deeper intellectual education, not only their families and the state.[86] Women authors were more centered on the particular features of each child, thus departing from the Lockean idea of children's minds as a tabula rasa. They championed the belief that learned mothers were the ones entitled to educate their children rather than men, as they spent more time with the children and better knew their intimate character.[87]

This approach was explicitly highlighted in the frontispiece of one of the many Spanish editions of *Les veillées du Château* (Tales of the castle). The author, Mme de Genlis, is portrayed at her desk and with a library behind her, a symbol of her erudition.[88] She is surrounded by boys and girls of different ages, who listen with amusement to her reading. However, Genlis is not only reading aloud but also "testing" her stories. As she explains in the prologue, before publishing the book she used to read these stories to children of different ages—the reason, she argues, that her stories were so successful among young audiences. In addition to being a good marketing strategy, this statement insinuates that women are able to understand children better than men. The Spanish male translator of Genlis's book stressed precisely this point. According to Fernando de Guilleman, Genlis was above all "a mother," and as such, she would succeed in educating the children much better than any educated man: "Her work is superior to any man's work, however wise and educated he might be: because he only writes by speculation, and if he gains some experience, he will never achieve what a mother would do when she educates her children, especially if she has talent and reflection."[89] Guilleman continues this point in the dedication, dedicating his translation to the Junta de Damas y mérito, as the "representative body of all the ladies of the kingdom."[90] In the earlier chapters of this book, I showed the importance of the Junta, which gathered together the most celebrated and wealthiest women in Spain. One of its missions was to promote the education of girls, and to that end, the Junta ran four schools in Madrid for poor girls, where in addition to textile work and religion, girls were taught reading, writing, and arithmetic. Since the ladies of the Junta were often represented as the "mothers of the nation," Guilleman was thus enhancing their maternal role as educators.

Female authors also had a broader opinion than male pedagogues about the goals of female education. The education of girls was often justified because of its social utility—it was thought that an educated woman would be a better companion to her husbands, a better housewife, and a better mother of future learned citizens. However, female authors, such as Josefa Amar y Borbón, Rita Caveda, and Ana Muñoz also praised women's education as a way to enhance their individual happiness.[91]

In addition, some women overtly recognized that women's education could also afford them political power of sorts. This was the opinion of one of the most well-known members of the Junta, the Marquise of Fuerte-Híjar, born María Lorenza de los Ríos (1768–1817). In 1795, the president of the Junta, this time the Countess of Truillas, proposed to create a commission on education to research why and in which subjects women should be educated. The Marquise of Fuerte-Híjar joined in and expounded an interesting reason for giving a broad education to elite girls: their future duty as advisors to men in positions of leadership. "And because they [women] should, in any situation in which they find themselves—whether as spouses, mothers, or friends—influence the workings of men in their civil and political functions, they would not be able to communicate useful ideas to the men if their education were careless," stated the marchioness.[92] Note that this latter term, "friend," or "amiga" in Spanish, appealed directly to the equality of the sexes, as enlightened friendship was only thought possible in a relationship among equals.[93] The learned women of the Junta were well aware that education could open ways for women to intervene in politics.

At the end of the century, Spanish reformers understood the scientific education of girls as key to national improvement. They turned to the new juvenile literature, which was regarded as a powerful tool of social and political reform, and launched a journal, the *Gazeta de los niños*, which would arguably allow them to reach a broader audience. Inspired by a French revolutionary journal, the *Gazeta* showed examples of fathers investing in the education of children of both sexes and of ways of "instructing [children] with delight" by using toys, domestic gadgets, and promenades in the countryside and the city. It is tempting to suggest that the analogies between the *Gazeta* and some elite pedagogical practices implicitly highlighted that the reformers needed the support of elite mothers to further their political agenda.

At the same time, high-ranking women, such as the Duchess of Osuna, her peers in the female patriotic association the Junta de damas, and female authors promoted a careful, broader education of elite girls. While some of these girls were able to share the lessons that their brothers' tutors gave at home, most of them likely read the juvenile literature of the time and played with pedagogical toys, scientific instruments, and natural collections. Among other reasons for investing in girls' education, learned mothers aimed to provide their daughters with the means of being politically and socially influential, either through their future marriages and/or social activities, and perhaps to contribute to the "public happiness": to be women improvers.

In the next chapter, we will delve into how these learned women actually practiced some of these lessons, more concretely in the area of rural economy.

5

Imagining a New Country

Women and Rural Economy

In eighteenth-century instructive literature, the gentlewoman who flees the city with her children to a ramshackle manor in order to make ends meet is a cherished character. Through her clever stewardship, she overcomes bankruptcy, makes barren fields grow, and educates virtuous children. In the popular *Les Veillées du Château* (Tales from the castle) by Mme de Genlis, for instance, the Marquise of Clemira leaves Paris when cutbacks are needed as her husband marches to war, to ensconce herself with her three children in her half-ruined estate in Champcery.[1] This was also the case of the protagonist in the *Économie rurale et domestique* (Rural and domestic economy) by Antoine-Agustine Parmentier, called "Marquise*" in the book, who retires to the countryside to save her family from ruin.[2] And the same is true for "the countess," a character who appears in the massive manual on agriculture and farming compiled by José Antonio Valcárcel, a member of the Valencia Economic Society.[3] Valcárcel included in the manual a letter supposedly written by a grateful countess, who, despite "being a woman raised in the big city," was able to turn the estate's poor sandy lands into gorgeous meadows thanks to the manuals of Valcárcel and other agronomists.[4]

These stories featured strong enlightened women in rewarding rural tasks. Authors presented ideal wives and mothers, as well as ideal environments that fruitfully respond to feminine care. There were at least two moral lessons these reformists authors intended to convey. First, that wives' duties included being a steward of the family's rural possessions, and second, that rural economy could be learned from manuals and journals. The countryside occupied a paradoxical place in imaginaries of the time. On the one hand, it was pictured as an ideal pastoral paradise, an Arcadia where happy virtuous citizens were to be educated close to natural goodness; on the other, it was considered a refuge of backwardness, superstition, and deep ignorance in need of urgent reform. In both images, the role of genteel women was considered key. As producers, consumers, and landowners, women were thought to be able to guide families and countries toward either ruin or success.[5]

According to Denis Diderot, rural economy was "the art of knowing all the useful and lucrative objects of the countryside, of procuring them, preserving them, and extracting the greatest advantage from them."[6] Consequently, the knowledge supposedly needed to succeed was huge, ranging from agriculture to gardening and botany, as well as domestic medicine, chemistry, arts, and architecture. It was expected that one learn a variety of skills, such as how to cultivate artificial meadows; preserve crops and fruits from pests, and livestock from diseases; make wine, oil, and dairy products; keep bees and cultivate silkworms; prepare hemp and flak; become expert in hunting, fishing, and falconry; manage stud farms, water, and forests; and manufacture a range of products such as stoneware, pottery, lime, brick, and iron.[7]

Rural economy was a bustling area of research at the time.[8] Agronomists, apothecaries, and natural philosophers were busy inquiring and writing on these everyday, mundane activities, which often took place in domestic settings.[9] Consider, for instance, the ideal country house depicted in figure 5.1, taken from one of the most important rural Spanish journals, *El Semanario de Agricultura y Artes dirigido a los párrocos* (The weekly journal of agriculture and arts devoted to parish priests). In the opinion of the *Semanario*, such a house should have rooms for distilling eau-de-vie, presses for making wine and oil; a room for the dairy; an oven for baking bread; stables for cows, goats, pigs, and horses; henhouses; a kitchen; a laundry room; and rooms for storage. To name but a few of the most famous authors in this genre, Antoine-Agustine Parmentier, the Abbé Baptiste François Rozier, the chemist Jean-Antoine Chaptal, and Benjamin Thomson, Count of Rumford all wrote on how to make cheese and lard, bread from different grains, cheap nutritious soups, soap from different types of fats and ashes,

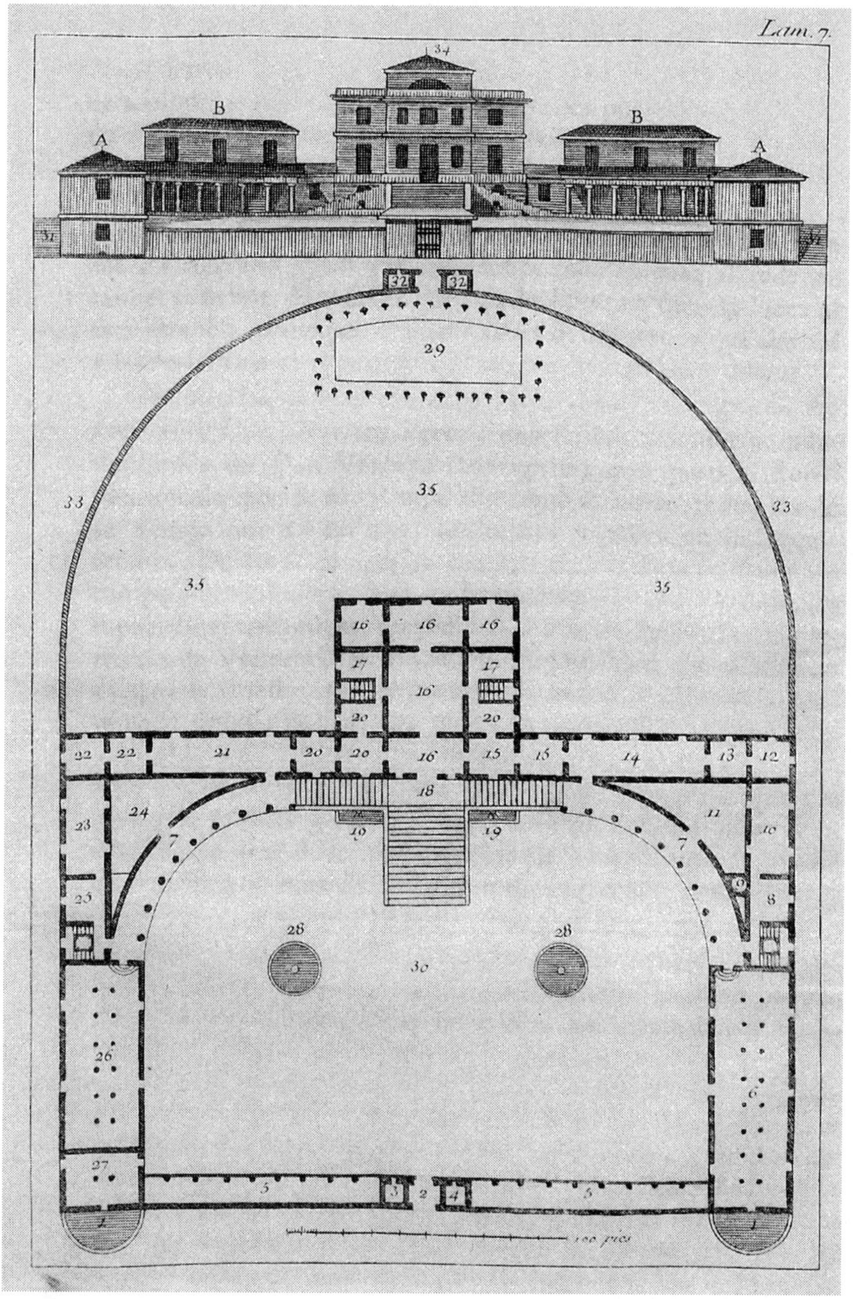

Figure 5.1. The ideal country house according to the *Semanario de Agricultura y Artes dirigido a los párrocos*. Biblioteca Nacional de España.

dyes, and spot cleaners.[10] These men believed that it was crucial for people to learn the "principles" behind how various things were done. Chaptal advised artisans: "Become better acquainted with the materials you work on . . . study the principles of your art, and you will be able to foresee, predict, and calculate every effect."[11] Journals, local societies, and rural schools were set up all over Europe to instruct peasants and craftsmen. As Parmentier put it, "only by means of popularization can science be made useful."[12]

Women were among reformists' primary targets. To be sure, countrywomen worked on agricultural tasks, but they also had prominent roles in food and textile matters. In his project for an agricultural school, Rozier listed all that peasant girls needed to learn: "sewing and knitting, getting the best from milk, cleaning and taking care of white cloth, working in the bakery and the kitchen, making wool and linen clothes, and lastly, according to season, [taking] on silkworms and spinning."[13] Reformers such as Rozier arguably intended to teach countrywomen the proper way to do all these chores. Yet this relationship also worked the other way around: these men sought out the experience of locals, whom they needed as informants and supporters. Just as in fields such as geology and botany, country folk and lay practitioners were instrumental here for communicating local variations and techniques and for cementing the credibility of savants, disciplines, and institutions.[14]

This chapter analyzes the dynamic interlinks between women as producers of knowledge and women as recipients of scientific discourses. It explores in particular how high-ranking women intervened in rural economy, and conversely how they were to be instructed. Women supposedly needed to know the latest advancements in their everyday chores (even if only to teach servants and daughters); at the same time, it was believed that they could contribute to general betterment through their observations, closeness to natural and domestic environments, and informal links with scientific networks. To explore these topics, we will focus on the pedagogical texts explicitly dedicated to women published by the *Semanario de Agricultura y Artes dirigido a los párrocos* (1797–1808). These texts are a good source for understanding how reformists addressed female audiences and envisioned their role in circulating knowledge. We will look at two of them in detail: the "Principios de botánica en cartas a una señora" (Elements of botany in letters to a lady) and the "Compendio de la chimica acomodado a la instrucción de las mujeres" (Compendium of chemistry suitable for the instruction of women). Written in language that is arguably plain and entertaining, the genre "science for women" had paradoxical functions, as we discussed in chapter 1. On the one hand, these texts contributed to the hierarchization and gendering of scientific knowledge. On the other, they make visible women's scientific interests, providing models of how women could create knowledge useful for the improvement of rural economy.[15]

In addition, we will examine the letters that women sent to the *Semanario* during the years 1797 and 1798, which will allow us to grasp how women situated themselves in this project of national renovation, and how they might have supported agronomists and economic philosophers.[16] As academic training was not required for one to be considered part of the learned community, genteel women could join scientific networks informally. They also used their social standing to wield influence as patrons, interlocutors, hostesses, and editors.[17] In the case of rural economy, only the well-off had the resources to cope with the pace and expenses of agricultural experiments, so some female landowners were able to ally with physiocrats and agronomists. As early as the years 1787 to 1789, in his journey to the south of France, Arthur Young quoted these gentlewomen, naming the Duchess d'Enville "la grande cultivatrice," while in his monumental *Agronomie et agronomes en France* (Agronomy and agronomists in France), André J. Bourde identified these women as *ameliatrices*—figures who embodied the enlightened landowner and tried out agricultural novelties on their lands.[18] More recently, Charles Loïc and Christine Théré have explored the networks of gentlewomen who propagated, in their salons, the *science nouvelle* of the physiocratic political economy and attended the "economic meetings" set up by the Marquise of Mirabeau.[19] In the case of Spain, we still need to uncover how agronomists might have used female aristocratic networks, but one way to begin is with the gardens of Spanish noblewomen. Often used to showcase political stances, the gardens of the gentry reflected changes in their social and political roles.[20] In the last part of the chapter we will thus explore how one of the most fashionable gardens in Spain, El Capricho (The caprice), owned by the Duchess of Benavente (or Osuna), might have stood for her commitment to agronomy and rural economy.

Setting up a Network of Improvers

In 1797 the clergyman Juan Antonio Melón and the civil servant Juan Bautista Virio embarked on a challenging pedagogical venture. They intended to use the network of parish priests to teach "rural economy" to Spanish farmers, including those in the colonies. It was not the first time that the clergy had been asked to offer their countrymen more than heavenly happiness. For instance, in 1784 and 1793 the Aragón Economic Society and the Royal Society of the Basque Provinces of Friends of the Country, respectively, published texts in which they encouraged rural clergymen to instruct peasants in the new agricultural techniques.[21] But Melón and Virio were far more ambitious. They would publish a national journal, the *Semanario de Agricultura y Artes dirigido a los párrocos*, with all the novelties it presented explained in an easy, pleasing way that priests could read to their parish after Sunday mass. In a country where "the ones who plough don't read, and the ones who read don't plough," Virio and Melón thought that priests would create a bridge for knowledge between literate and illiterate people.[22]

The *Semanario* was conceived in a climate of a broken national economy, but also of relative openness and faith in education. The war, first with Britain and then with revolutionary France (1792–1795), had left the country nearly bankrupt. Yet the political situation was more permissive than before in intellectual and religious terms and reformist-oriented in its policies.[23] The Inquisition had been limited in its scope and formerly banned newspapers were being reopened—on the condition that matters concerning religion and monarchic government were not discussed. Books under inquisitorial censorship—including the *Encyclopédie méthodique*—were now allowed to be published, and the translation of foreign works accelerated.[24]

Historians have stressed the Spanish obsession with foreign technological novelties, which were seen as a panacea for national problems. As Richard Herr states, "in the eyes of the ministers, one of the greatest needs of Spanish craftsmen and farmers was knowledge of advances being achieved abroad."[25] In this regard, the editors of the *Semanario* were well equipped. Both Juan Antonio Melón (1748–1843) and Juan Bautista Virio (1753–1837) were close to the "afrancesados" (the name used to design those supposedly close to French philosophical ideas and costumes), to the point that they would back José I Bonaparte in 1808 and had to go into exile afterwards.[26] They had also traveled widely. Appointed to the Spanish embassy in London between 1783 and 1786, Virio negotiated the mercantile treaties between Britain and Spain. In 1790 he was sent to investigate how European countries organized their finances, agriculture, and industry. Over a period of four years, Virio journeyed across France, Switzerland, and Germany.

He spent months in Lyon perusing the silk manufactories, visited glass and lingerie factories in Tirol and Bohemia, and observed agricultural practices in Brunswick, Göttingen, and Austria. When he returned to Madrid five years later, Virio proposed to the prime minister, Manuel Godoy, a plan to expand education on political economy (*Plan de educación económico-política*), in which he suggested that clergymen should be able to instruct peasants. For his part, the clergyman Melón traveled during the 1780s through France and England, and in 1798, was appointed to coordinate all Spanish economic societies. In addition to being an enthusiast of botany and the new empirical sciences, Melón navigated well in polite society.[27] All these foreign experiences were to be very useful to the *Semanario*.[28]

In addition to its usefulness, the *Semanario* seemed to be a lucrative business. The proceeds from potential buyers—all parishes in Spain and their colonies—amounted to around 31,000 *reales de vellón*. The government sent letters to all the Spanish bishops encouraging them to call out the rank and file of the clergy. More than twenty-two thousand copies of the prospectus were distributed, and three thousand copies of the first number of the *Semanario* were printed (the normal edition was five hundred copies). However, Melón and Virio had in mind a broader audience than clergymen and their peasants, as is apparent from the two different editions of the *Semanario* (figure 5.2).[29] Both frontispieces conveyed the same message on the relationship between agriculture and prosperity—as represented by the horn of abundance filled with fruits and rural tools. But while one was signed by the prestigious engraver Luis Paret and showed a large-windowed country house with the iconic lightening rod, the other engraving was anonymous and showed a humble peasant house. Although these different editions may have been due to external contingencies, they might also have reflected the wish of editors to appeal to higher classes. Beginning on January 7, 1797, it was possible for anyone to buy a copy every Thursday at the local post office, bishopric, foundling home, or hospital. An annual subscription cost 75 *reales* in Madrid, 114 in the Spanish provinces, and 220 in America (note that laborers earned between 5 and 11 *reales* a day).

In their headquarters in Madrid, the editors fully prepared themselves to completely fill the sixteen pages of the journal.[30] They bought four leather desks; stored seeds, which they planned to share with subscribers; acquired an aerometer—an instrument for measuring the density of liquids; bought "machine models" (it is unclear whether these were just engravings or scale

N.º 1.

SEMANARIO

DE AGRICULTURA Y ARTES

DIRIGIDO Á LOS PÁRROCOS.

Del estudio de la agricultura.

Ya hace mas de diez y siete siglos, que el sabio agricultor español Columela decia á los Romanos de su tiempo, „que no debian atribuir sus cortas cosechas á la esterilidad del pais, sino al descuido con que miraban á la agricultura, abandonada á mercenarios, contra el exemplo de sus mayores, que se gloriaban de exercerla, al mismo tiempo que desempeñaban los cargos mas importantes de la república, pasando, tal vez, desde el carro triunfal, con que la pátria premiaba sus hazañas, á empuñar, cubiertos de laureles, la esteva en sus cortos campos, que, como si lo agradeciesen, daban abundantes cosechas. Admirábase mucho del cuidado con que se buscaba un maestro de eloqüencia, de música, de bayle, y de todas las demas enseñanzas, quando la agricul-

a

Figure 5.2. The first page of the *Semanario de Agricultura y Artes dirigido a los párrocos*. Notice the frontispiece designed by Luis Paret. Biblioteca Nacional de España.

models); and invested in a new press to engrave "instruments, machines, fruits and rural buildings."[31] They assembled a list of rural journals published in Europe and subscribed to many of them. The *Semanario* would translate many articles from French and English agricultural journals, such the *Journal des Arts et Manufactures* (Journal of arts and manufacturing), the *Journal d'Observations sur la Physique, sur l'Histoire Naturelle et sur les Arts et Métiers* (Journal of observations in physics, natural history and arts and crafts), the *Annales de Chimie* (Journal of chemistry), the *Feuille du Cultivateur* (Leaves of a farmer), and the *British Encyclopaedia*. The editors were also able to purchase all 152 volumes of the *Bibliothèque universelle des dames* (Universal library for ladies), published from 1788 to 1797, which provided them material for their articles on domestic economy.

Often, these articles included recipes reminiscent of those in the *Books of Secrets* of the past centuries, for instance, for how to wash linen, or how to purify honey to be used as sugar.[32] However, in contrast to this earlier model, the articles showcased chemical principles.[33] For example, Chaptal's article about how to make soap in factories and private households began with a section called "Theory of This Art," in which he explained the formation of alkalis. Upon illustrating how to obtain cream from milk, Rozier explained in detail the chemical components of milk, while in Guyton de Morveau's recipe for purifying the air, readers learned about how acid destroyed alkaline miasmas.[34]

Yet in addition to instructing its readers, one of the main aims of the *Semanario* was to be an open forum for the Spanish rural community. Melón and Virio expected to entice the curiosity of their readers and motivate them to engage in experimenting and sharing their observations:[35] "This journal will be a hub for useful news to our farmers . . . it will rise the curiosity of the landowner and the diligent craftsman to carry out the proposed trials (*experimentos*) and to report to the *Semanario,* which will publish the results for the common good."[36] As a useful strategy to promote popular involvement, posting letters to the *Semanario* was free. Also, in order to publicly recognize the merits of contributors, the editors listed, in the index of authors, known figures in sciences such as Linnaeus and Lavoisier alongside laypeople, such as the peasant Cándida and her method of preventing smallpox, and one Matilde G. Sendin, who had submitted a chemical translation.

In summary, the *Semanario* aimed to pitch a national project of education and improvement to a large part of society; and it intended to create committed networks of "rural economists" engaged in experimenting, making the most of natural resources, and circulating their knowledge. How did the editors expect to accomplish this? How did they expect the journal to be read? We have some hints in the letters that the parish priests sent to the editors. The performance of the communal readings of the *Semanario* resembled Catholic mass, with the written word either coming from God (the Bible) or from men in sciences (the *Semanario*), explained by the mediating role of priests. In addition to meeting in church porticos and rectors' houses, the people might have met in the local town hall, as explained in a letter published in the *Semanario* from the clergyman of Villamayor de Santiago.[37]

The fact that the editors published this letter in the *Semanario* means that they endorsed his paternalistic, even reverential approach, and thus it merits describing how Villamayor planned to read the *Semanario* in some detail. According to the plan, on the first Sunday of the month the church bell would toll at four o'clock in summer and at two o'clock in winter, summoning all the people who worked in the fields and the authorities—the town's two mayors, the vicar, the confessor of the convent, the doctor, and the apothecary—to the Council Hall. The parish archive would keep exemplars of the *Semanario* and the reports about any conducted trials for later uses. Setting the time suggests the intention of a regular practice; the bell signals the event being important in the life of the community; and the milieu linked the communal readings to issues of governance. The presence of the doctor and the apothecary suggests that the event deserved scholarly attention, while the intention of recording trials indicates the intention of contributing to the "economic history of the province," as the former minister and instigator of economic societies Pedro Rodríguez de Campomanes noted.[38]

Because of the letters that the same priest sent some months later, we know that the plan seemed to work, as the people of Villamayor endorsed the priest in his tests of making soaps from different ingredients. Some families provided kitchen ashes and olive oil leftovers, and some laundresses tested the greyish soap that the priest eventually produced.[39] According to the priest, the new way of making soap benefited all: "All the village is happy with this new invention, which has already been repeated in four households."[40]

Beyond ecclesiastic authorities, lay practitioners and in particular women also sent their letters to the *Semanario*. Who were these women, how did they fashion themselves, and how did they seek to contribute to the public good?

The Women Improvers of the *Semanario*

During the first two years of its existence, the *Semanario* was a success. More than 25 percent of its content came from readers of all walks of life.[41] Clergymen, as we have seen, but also gentlemen, members of the military, craftsmen, fellows of economic societies, and women put pen to paper.[42] Their letters give hints about their interests and about how they fashioned their identities as improvers. However, we need to consider that published letters could have been "amended" by the editors or even forged, as letters were often vehicles to buttress the social, political, and moral purposes of a journal. Indeed, the editors of the *Semanario* probably authored some of the letters—for instance, a suspiciously well-written one signed by "a peasant" that was published after a sharp, polemic discussion on the distribution of lands, in which a peasant heartily defended the large land properties of the Duchess of Alba. Or a dialogue between a priest and a peasant, supposedly an eyewitness report, in which the priest convinced the peasant to read the *Semanario*.[43] Moreover, in the *Semanario* the letters had to pass two filters before being published. They first had to be approved by the censor Pablo Forner, and then by Prime Minister Godoy, so of course they contained nothing subversive or in disagreement with the government's policies.[44]

Still, it is precisely because of the resulting alignment of these letters with the hegemonic view of the time that readers' letters are so helpful to the historian. Women's letters cast light on which topics and practices were gendered, which strategies women used to legitimate themselves, which intellectual networks they belonged to, and which female models were thought appropriate. Letters authored by women in scientific journals were certainly not as frequent as men's; for this reason, they were often showcased as examples of edifying behavior to other women. And yet beyond this exemplarity and compliance with norms, women's letters allow us also to grasp women's initiatives, complaints, and voices.

The letters authored by women in the *Semanario* are scarce. For instance, out of a total of forty readers' letters in the first part of 1798, Elisabel Larriba only found two signed by women.[45] However, women were often present in men's letters and other articles. We can see them working behind the scenes, busy trying and tweaking recipes as in the case of the soap trials in Villamayor, in herbal gardens, in chemical laboratories, and at their desks with pen and paper. Consider, for instance, a letter on the uses of cherry tree sap. It was signed by a male "subscriber from Lerida."[46] He carefully fashioned himself as "a curious one" and recounted the following story. He was wondering whether he could use the sap of fruit trees as a substitute for imported gum Arabic, when María Biosca told him about her discovery that cherry tree sap could be used to starch clothing in the same way as gum Arabic. The subscriber from Lerida then tested María Biosca's recipe along with his wife and "it worked perfectly." But as he was a "curious one," he systematically tested the sap of other fruit trees, in their wild and grafted varieties, and observed the structure of their sap with a microscope. He was then able to determine that the wild variety of cherry tree worked better than any other. However, he accorded María Biosca some merit: "But in spite of this [of not specifying the right variety] Biosca deserves praise for her industrious invention, which I have tried to spread in this city with some success. Let us hope that it is also introduced to the rest of Spain! In this way, not so much money will leave the kingdom."[47] The letter highlighted the difference between its male author, featuring him as a proper researcher capable of systematically comparing different botanical varieties and even using the microscope, and his informant María Biosca, who was presented as merely handing him the recipe without carrying out any further inquiry. And yet the letter represented the type of women contemporaries would call "economic."[48] Biosca accomplished the cherished goal of political economists: profiting from natural resources in novel, unexpected ways. Saving gum Arabic, Biosca saves on the family budget and reduces imports in the Spanish trade balance. In the few lines of the letter, the readers of the *Semanario* saw a woman taking care of her clothes, looking after her garden, and storing a provision of sap in the spring to be used throughout the year.

Biosca represented the economic woman—an ingenious, diligent, and thrifty woman capable of doing without foreign products. Such was also the case of Maria Morales, a woman from the village of Lucillos near Madrid. Morales put pen to paper to explain how to make a "delicious meal" with a plant that was normally used to feed animals, the grass pea or chickling pea, *almorta* in Spanish. In her four-page letter, Morales provided detailed information about how to cultivate the grass pea. She begins her letter with the phrase: "As soon as I heard that the *Semanario* encouraged the use of polenta." In addition to reminding us that women did not need to be subscribers to have access to journals, this also shows how women may have collaborated with improvers' research agenda—in this case, by incorporating new crops into their household's diet.[49]

Nevertheless, in the *Semanario* female economic practices extended beyond domestic realms to encompass entrepreneurial women, translators, and diarists. A male traveler who passed by Aldea del Río explained in detail the success of a cloth workshop run by a woman of the village, in the way reformist enterprises ought to be run, that is, commissioning spinning and other tasks to women in private households instead of gathering them in large manufactures.[50] The husband of another woman, Doña María Belaunde, noted that his wife designed a method for making better use of waste paper in the family paper mill.[51] The wife of the Spanish minister in The Hague, who signed herself María Cuenca, published her observations about the Swedish country gentry, who engaged themselves in the "pleasant tasks of cultivating gardens and lands" and instructing peasants, who in turn had "beautiful and hygienic houses."[52]

As noted above, the behavior of these women was held up as an example to be followed. For instance, the editors praised Cuenca for bringing news and encouraged Spanish travelers to do the same: "It would be very appreciated for national travelers to behave . . . with the zeal that distinguish the Lady who has sent the letter . . . whose observations in Sweden and her journey to this country from La Haya have proven worthy to us of being published."[53] In addition to women endorsing reformers policies, the *Semanario* also contained examples of gentlewomen circulating new techniques and news. It reported, for instance, that the female society Las Señoras de las Cárceles that we encountered in chapter 3 were "well aware of what has been written about the purification of air," and that they were now using the method of nitric vapors to increase the content of oxygen in Madrid's jails. The journal explained how the Duchess of Infantado and the Duchess of Alba installed a lightning rod in their home, a novelty at the time. Moreover, a long article on the Duchess of Alba exemplified well how the editors expected gentlewomen to promote rural knowledge. The duchess not only funded the translation of Rozier's massive sixteen-volume dictionary on agriculture; on her lands, she also invested in rural machinery, carried out agricultural trials (though it is not specified how), and ordered her foreman to instruct workmen and peasants "with gentleness and affability."[54]

Finally, some women also collaborated in translating scientific articles for the *Semanario*. For instance, one Matilda G. Sendin translated a long article from French, published in three consecutive issues, on how to produce saltpeter to make gunpower. It is interesting to note how she negotiated this incursion into a supposedly male venue. She apologized for explaining "a damned invention that men discovered for the madness of killing themselves" and noted that she planned shortly to translate an article on types of yarn.[55]

A contrary case is that of María Antonia Gutiérrez Bueno (1781–1874), daughter of the royal apothecary and professor of chemistry Pedro Gutiérrez Bueno, and his first wife, Mariana Ahoiz, who lived in France after their marriage and inherited her father's apothecary in 1822.[56] Rather than apologizing for writing on an arguably male topic, M. A. Gutiérrez Bueno proudly showed herself to be an expert on chemical knowledge and acquainted with chemical laboratories. In 1800 she published in the *Semanario* a translation from the French *Décade philosophique* on the medicinal properties of acetic ether and a method for its distillation in a sand bath.[57] Her article appeared directly before one written by her father, perhaps as a way of telling readers where her chemical wisdom came from. A year later, she published a translation from the *Journal des Arts et Manufactures*. The six-page article described a method for synthesizing ammoniac salt—a combination of the volatile alkali ammonia with muriatic acid that was made by burning organic material with common salt in special ovens.[58] Gutiérrez Bueno included a large footnote where she clarified some crucial aspects of the procedure: she suggested the ideal clay vessel's shape for collecting the salt, specifying the exact quantities of raw material needed and the length of time they should be heated.[59] Moreover, she described another method for obtaining the salt, which she had witnessed at the Madrid Royal Chemical Laboratory in 1792. She even recommended that factories to produce ammoniac salt be established in Spain. Years later, under the pseudonym Eugenio Ortazán y Brunet (formed from all the letters of her name), she would publish on the cholera epidemics that erupted during the first decades of the nineteenth century and begin a biographic dictionary of famous women.[60]

Gutiérrez Bueno was not the only woman who appeared in the *Semanario* discussing an arguably male topic. The *Semanario* also published the experiments of the French Marie-Armande-Jeanne Gacon-Dufour with chickens to improve their breeding in wintertime.[61] Mme Gacon-Dufour was a prolific author, well-known in the scientific circles of Paris as well as among women readers. Her manuals of household management had been edited several times, while her research on food preparation and preservation, and on mechanical arts and apparatuses to avoid using too much combustible material, had been published in economic publications such as the *Bibliothèque Physico-Économique*.[62] A footnote to her article that was translated in the *Semanario* advised that her conclusions were not accepted by everyone. This note was made in the same way as in many other scientific articles that were discussed in the journal, leading me to suggest that, rather than diminishing her, it is a sign that she was afforded the same respect as other male scientists.[63]

In summary, the way that women appeared in the *Semanario* showed how an enlightened female population could contribute to the improvement of the nation. But in order to do so, women needed to be taught in the new sciences. The editors published specific pieces dedicated to women's scientific education. In 1801 the *Semanario* launched four manuals to teach women the elements of chemistry, botany, agriculture, and natural history. According to the *Semanario*, women should be familiar with this knowledge: "To the *Seminario*, we owe the publication of several good elementary manuals of chemistry, botany, agriculture, and part of natural history, written in an easy style and accommodated to the understanding of women . . . to inspire the new generations with the important knowledge that until now has unfortunately been too far from our education."[64] The genre of "science for women" was also thought to be suitable for the uneducated classes, as authors presupposed that lower-class men shared similar intellectual features as women. The female reader supposedly lacked any previous knowledge on the topic, was ignorant of the scientific nomenclature, and had no mathematical skills. She was thought to be inconstant and inattentive, necessitating that the texts be entertaining and pleasant to read.[65]

The four manuals devoted to women—the "Compendio de la chimica acomodado a la instrucción de las mujeres" (Compendium on chemistry suitable for the instruction of women), "Principios de botánica en cartas a una señora," (Principles of botany in letters to a lady), the "Elementos de agricultura" (Principles on agriculture), and the "Elementos de historia natural en cartas a una Señora" (Principles of natural history in letters to a lady)—followed the same structure: a learned man instructed a clever woman, called "C.," by correspondence.[66] However, because we only have the letters of the male teacher, we can only know the response of Lady C. through his commentaries. The series began in 1801; over a period of almost two months the *Semanario* published a piece on chemistry every week. In September 1802, when C. was done with chemistry, she declared she wanted to learn botany.[67] In 1803, she asked to be trained in the principles of agriculture.[68] Finally, in 1805, she requested that her by-then exhausted friend illuminate her in the principles of natural history.[69]

In the following, we will examine the texts on chemistry and botany. The one on chemistry was an abridged translation of the Italian best-seller *Chimica per le donne* (Chemistry for women, 1796), while the botanical text was an adaptation of the lectures given by the director of Madrid Royal Botanical Garden Antonio José Cavanilles in 1801.[70] We will see the transformations that these texts underwent, which were probably due to Melón's editing and translation. In particular, we will see how the new texts framed the involvement of women and laypeople.

Women of Chemistry

The "Compendio de la chimica acomodado a la instrucción de las mujeres" begins in the following manner: "You say, kind C., that you would like to learn chemistry because it is a fashionable science, and as following the fashion in the study, it will not be seen as the eighth cardinal sin."[71] If we leave aside the ironic criticism of how women were educated, "kind C." certainly had a point: during the 1790s chemistry was in fashion in Spain. Not only were chemistry chairs being funded at medical faculties, apothecaries' schools, and national laboratories such as those in Madrid and in Segovia but chemistry was also being supported by military institutions, economic societies, and schools for craftsmen, including the School of Artillery in Segovia, the economic societies of País Vasco, Aragón, and Valencia, and the Junta de Comerç in Barcelona.[72] In arts such as metallurgy, the manufacture of dyes, and food processing, chemistry was also seen as key. The quick expansion of chemistry was favored by a strong policy of fellowships for studying abroad, appointing foreign chemists to translate chemical manuals and dictionaries, and setting up specialized journals.[73]

However, beyond these mostly male venues, female audiences might have become acquainted with new knowledge in chemistry through public spectacles and itinerant courses that showcased scientific instruments, demonstrated natural laws, and exemplified the usefulness of chemistry.[74] In one such spectacle staged in Madrid in 1799, the showman promised to exhibit "the way in which mineral waters can be found, how wines are falsified, and how to put two liquids together without mixing one with the other."[75] The announcement of the course given by Francisco Bienvenu in October 1797 clearly revealed that he had a keen female audience: "Don Francisco Bienvenu, professor of physics, to please the public who wishes to know whether ladies can assist the course that will begin next Monday the ninth, has the honor of announcing that among the subscribers there are already several ladies of this court, as is usual in Paris and other cities of Europe."[76] Bienvenu (1758–1831) was an instrument maker who had previously taught physics to a high aristocratic Parisian clientele.[77] After the French Revolution compelled him to change careers, he performed shows all over Europe. In his Spanish spectacles and courses, he used a great variety of scientific instruments, including solar microscopes, the Volta eudiometer for the purification of air, and electrical machines of different types. Other chemistry courses in which women may have participated include those offered by the itinerant lecturer Italian Giuseppe Pinetti, who performed in Madrid during the 1790s. Pinetti also offered private tutorials, as did others such as the mineralogist Christian Herrgen (1760–1816) between 1798 and 1803, or the chemist Pedro Gutiérrez Bueno in his laboratory, although we do not know if women attended.[78] In all cases, Spanish aristocratic ladies were probably well aware of the fashionable Paris and London chemistry courses, in which women comprised half of the public.[79]

In short, when the *Semanario* published its chemistry for women, chemistry was already considered a useful science and had an eager lay audience. In fact, at the same time that the *Semanario* published the translation of Compagnoni's *Chimica per le donne*, another one appeared in Barcelona by Josef Antonio Sabater.[80] However, the two translations were very different. While Sabater was faithful to Compagnoni's text, the *Semanario* changed both its content and style.[81] Interestingly, Compagnoni also based his text on a chemical encyclopedia. The chain of transformations from the chemical encyclopedia to Compagnoni to Sabater and to the *Semanario* casts light on the differing political and didactic purposes of these texts.

According to Compagnoni, his *Chimica per le donne* was born in a Venetian *spezeria* after a bet.[82] Drinking with his chemical friend Dandolo and the publisher Alessandro Pepolo, Dandolo joked about Compagnoni's inability to understand chemistry. Compagnoni responded that in one month he would bring out a chemical manual.[83] He would adapt Dandolo's chemical encyclopedia, *Fondamenti delle scienze fisico-chimica*, which had just been published by Peppolo.[84] The *Fondamenti* aimed to show the "philosophical-chemical picture of the universe"; in addition to being alphabetically ordered, the articles were classified in three groups, or tables.[85] The "tavola prima" listed the "essential" articles, that is, the history of chemistry, the attractive forces that drive chemical reactions, the definitions of simple and compound bodies (air, water), the elements (light, heat, oxygen, nitrogen, phosphorus, and the rest), combustion, and the formation of oxides, acids, and salts. The "tavola secunda" contained articles that explained organic processes: fermentation, putrefaction, respiration, digestion, production of animal secretions (urine, tears, semen), and animal sensibility and irritability. Finally, the "tavola terza" listed the meteorological articles (the formation of clouds, volcanoes, and earthquakes). Compagnoni's *Chimica per le donne* faithfully followed the *Fondamenti*. It was organized as a series of letters of a male tutor to a countess who wished to learn chemistry. The letters were ordered in three groups that followed the three *tavolas*; however, Compagnoni embellished his text with gallant digressions.[86] But it seems that he soon got tired, and the light-hearted spirit of the first part evidently flagged as the book progressed.[87] Moreover, he included a fourth section ("L'Appendice"), copied from a mathematical treatise. However, in spite of all its flaws, the *Chimica* was a success. It was thought of as being "simple, easy, and with doctrinal precision" and ran several editions.[88]

So how did the two Spanish translations compare? Sabater's translation followed Compagnoni to the letter, including the mathematical fourth part. The *Semanario*, by contrast, not only skipped that section but also removed all references to aristocratically shaped knowledge, reduced the gendered language, and dispensed with all flirtation. While Compagnoni dedicated his book to an "educated and distinguished gentlewoman," the *Semanario* dedicated it to "a learned lady." Instead of a countess, the protagonist of the *Semanario* was a lady who was always addressed as "dear C." or "kind C." Instead of the respectful "Vuesa Merced" that Sabater used, the philosopher of the *Semanario* addressed its lady with the familiar "tu" (you). Moreover, the *Semanario* cut out all jokes about the aristocracy or gender. For instance, Compagnoni remarked that women were excellent teachers of chemistry because they used it to "transform men's hearts in their stills." The entire passage was deleted from the *Semanario*, which also cut out the famous passage that encouraged readers to color all their dresses, furniture, jewels, trousseaux, and even butlers black in order to facilitate the absorption of light in winter, probably because the journal's expected readers would have own none or few of these fancy items.[89] For the *Semanario*, the model of a woman interested in chemistry was no longer exclusively a gentlewoman, nor was the language of seduction and polite gallantry pertinent.[90]

Moreover, the *Semanario* seemed to consider the polemics on the new chemistry settled. From the beginning, Compagnoni set out a battle between modern and old chemists. In fact, he suggested that one of the aims of the book was to give readers the means to judge between the "old" and "modern" chemistry. He compared Lavoisier to Galileo and to Newton, whom he saw as precipitating a revolution: "And Lavoisier was in chemistry what Galileo was two hundred years before in physics—destroying at once old formulas, the majority of which were false, inexact, absurd, or imperfect, and inventing a new language corresponding to the new principles."[91] Compagnoni and his translator Sabater pictured chemistry as a battle of a brave Lavoisier against ignorance. By contrast, the text in the *Semanario* erased all polemics. It kept the history of chemistry to the essentials and excluded any contrary argument. This absence of history of chemistry could be interpreted as the *Semanario* considering the polemic between the old chemistry and the new defended by Lavoisier already settled, in favor of Lavoisier.[92] However, what it did consider necessary was to focus the attention of the reader on important concepts, highlighting them in italics.

In any case, the *Semanario* provided their readers with actual, deep chemical knowledge that played down class and gender differences. The knowledge explained in the *Semanario* was that which was deemed necessary to understand the workings of the world and the body. It also served to further the understanding of everyday material transformations, in accordance with Chaptal's idea, mentioned at the outset of this chapter, that craftsmen should be acquainted with scientific principles in order to improve their production.

Botany for Women

This same idea of being useful is present in the "Principios de botánica en cartas a una señora," which the *Semanario* published after the "Compendio de la chimica acomodado a la instrucción de las mujeres." However, in the case of these "Letters to a Lady"—as I will now refer to them in short—the will of instructing women was forcefully stated, as well as the intention of extending knowledge of botany among them. By 1800 botany was thought to be perfect for the education of girls—an idea that was reinforced by Rousseau's botanical manual and its subsequent translations: "At any age, the study of nature abates the taste for frivolous amusements, subdues the tumults of passion and bestows upon the mind salutary nourishment by filling it with a subject most worthy of its contemplation."[93] Despite its benefits for female minds, botany was of course also seen as a serious discipline, considered indispensable for improving national wealth. Botanical gardens competed in adapting foreign plants such as textile plants and timbers, new crops for nourishing the population, and fashionable commodities such as tee, sugar, and coffee.[94] In fact, the original source of the botanical treatise of the *Semanario* was the lessons that the director of the Spanish Royal Botanical Garden Antonio Joseph Cavanilles (1754–1804) gave in 1801, in which he included a section on the elemental principles of botany (*Principios elementales de botánica*).[95] However, the *Principios elementales de botánica* and the "Letters to a Lady" differed in significant ways, beginning with how they organized their information, in their metaphors, and in the style they used. But they also differed in that in the "Letters to a Lady," botanizing was portrayed as embedded in contemporary practices of sociability.

The "Letters to a Lady" were organized as a correspondence between the same Lady C. who appeared in the chemistry lessons and her male friend. Here, Lady C. wants to teach botany to her little daughter Matilda, for which she herself needs instruction first. In the initial letter, her friend happily agrees, pointing out the moral benefits of botany and giving some advice on how to begin. He encourages Lady C. to collect flowers and observe them. He lists the necessary instruments for her to do so: a magnifying glass, a small pair of scissors, tweezers, and a narrow needle. Botanical treatises for ladies, such as those of Rousseau or Priscilla Wakefield, began with flowers, normally with the "eminently beautiful" lily, in which the sexual organs were easy to distinguish.[96] As is well known, the Linnaean classification was based on the number and distribution of a plant's sexual organs. To explain fertilization, the friend of lady C. uses the cherished metaphors of the marriages of plants, in which the calix was the nuptial bed and the pistils and ovaries the groom (frequently grooms) and bride.[97]

The educated reader probably knew Erasmus Darwin's famous poem "The Loves of the Plants" (1789), a best seller in Britain that had been recently translated into Spanish by José Viera y Clavijo, tutor of the children of the Duke of Infantado.[98]

The next letters are very technical and follow Cavanilles to a tee but with a twist. The tutor exhaustively explains to Lady C. the anatomy of every part of the flower (calyx, corollas, stamens, and pistils). Well aware of the dryness of the topic, he uses two pedagogical aids. First, he separates essential from descriptive information with two different kinds of typography. The second letter, for example, begins by defining "calyx." At the bottom of the page in a smaller font, the tutor describes the classifications of the calyx according to its traits. This same structure serves to explain fruits, seeds, roots, stems, leaves, and buds. Second, he frequently intersperses encouraging messages: "I beg you dear C. not to lose the hope of enjoying the anatomy and description of plants"; "I am encouraged to see, oh virtuous C., the constancy with which you admire the flowers"; "You are right to say that my letters are too dry."[99] But he suggests that Lady C. just look at the tables whenever she has a flower in her hand and not attempt to learn them by heart.[100]

This of course was not the case in the *Principios elementales* of Cavanilles. Cavanilles did not typographically distinguish information; on the contrary, he unapologetically inserted unending lists into the text. Moreover, following other male-fashioned Spanish botanical treatises, professor Cavanilles completely avoided the gendered metaphors of beds and marriages.[101] Instead, he provided his readers, the male students in the botanical garden, with rich accounts of the history, controversies, and problems surrounding many different botanical issues and exhaustively argued his position against or for the views of eminent botanists with hundreds of botanical examples.

Finally, after Lady C. has learned the entire "botany alphabet," as her instructor friend puts it, she begins to botanize according to the Linnaean system, of which the *Seminario* included an engraving.[102] She was also advised to read the classical volumes of Antonio Palau y Verdera (1734–1793) (the translation of Linnaeus's *Species plantarum* adapted to the Spanish varieties).[103] In the last letters, the male tutor encourages Lady C. to make an herbarium, the "Flora's library": "Spend the day with your family in the country and look for plants and place them in their classes, orders, species; I suppose that you will have with you magnifying glasses and the rest of the instruments, and some paper booklets to put the plants in with their flowers."[104]

Rather than an isolated domestic pursuit for women, the manner in which the "Letters to a Lady" pictured the study of botany was fully embedded in contemporary practices of sociability. Ingeniously mixing fiction with reality, the letters are awash of anecdotes that cast light on contemporary practices among learned elites. For instance, while botanizing with her instructor, Lady C. meets people from the closest circles of Cavanilles and Melón, such as the writer Leandro Fernandez de Moratín (1760–1828), Francisco de Goya, and the Duchess of Salm-Salm. In one of the first letters, her instructor remembers an encounter between Lady C. and Goya in the Escorial —an occasion for favorably comparing nature's spectacle with artificial paintings.[105] Lady C. also visits the garden of one Inarco Celenio, which was the pseudonym of Moratín. On another occasion, Lady C. converses about botany in the "beautiful garden" of the Duchess of Salm-Salm, and even lectured a young man on the topic.[106] Finally, Lady C. encountered the famous botanist Luis Née botanizing in Vallecas, in the outskirts of the capital. Née (1734–1807) had joined the five-year expedition of Malaspina around the world (1789–1794) and was able to gather more than twelve thousand specimens. Upon arriving in Spain, he worked with Cavanilles and his collection formed the nucleus of the Botanical Garden Herbarium. In the "Letters to a Lady," Née was described as "venerable old man, of white unruly head, sweet countenance, rosy color, active and diligent, who "luckier than Antonio Pineda [the other botanist of the expedition, who died in Philippines], has come after circumnavigating the world to give us news about plants."[107]

"Letters to a Lady" also included bitter complaints about the limited educational opportunities available to Spanish women, who were not able attend the courses in the Royal Botanical Garden, in comparison to their foreign peers: "Among us, women seemed exiled from these gatherings."[108] The last letter ended with a plea not only for women to learn botany but for the whole population to do so as well, in order to overcome the national crisis that made Spain economically dependent on other countries: "If more people applied themselves to acquire this knowledge with the same constancy as you do, it is unspeakable what agriculture, industry, medicine, natural history . . . would be able to advance, but if we cannot distinguish the precious production upon which we perhaps trod, if we cannot even recognize the species of wheat . . . we will live ignorant, poor, and condemned by all the nations that have learned to get from the natural sciences the wealth and power with which they threaten to subjugate others."[109]

One may wonder about the degree to which the eager-to-learn "kind C." might have been inspired by a real woman of flesh and bones. One possible candidate was Melón's own niece, Luisa Gómez Carabaño. She engaged in the study of botany and later would translate a botanical manual.[110] Alternatively, the model that Melón may have had in mind may have been the aforementioned María Antonia Gutiérrez Bueno, with whom he had a close friendship. The fact that these manuals for women were published along with other texts addressed to women, such as Parmentier's on how to make bread or cheese or Chaptal's methods for making soap, supports the thesis that the editors sought to recruit women in their project of modernizing the Spanish rural economy.

But what was the opinion of those to whom this instruction was targeted in the first place? What did peasants and lower classes think about the communal readings of the *Semanario*? Were these texts for women thought to be interesting, entertaining, and useful? Did the *Semanario* change the lives of peasants and artisans in any way?

Peasants' Voices

It is often difficult to find hints of how lower classes thought, but in this case, the needs of marketing have provided us with a survey that gives some indication. In 1802 the sales of the *Semanario* were not as expected and Melón attempted to get the government to make its purchase mandatory for parish priests. However, before agreeing to do so, the Consejo de Castilla commissioned the Bishop of Barbastro, an enthusiastic supporter of the *Semanario*, to carry out a survey to figure out how effective the *Semanario* had in fact been in his bishopric. After interviewing all his priests and some peasants and painstakingly perusing the parish books in which the trials were supposedly registered, the bishop submitted a twenty-two-page report. According to his findings, some peasants had succeeded in cultivating new seeds (corn, beetroot, *esparraceta*, and *alfaz*); several villages cultivated potatoes to feed humans and hogs; and others carried out "small trials" with seeds (washing them with alkaline solutions and ashes before planting them) to see whether these methods effectively protected the crops from pests. Some peasants experimented with grafting, while others reported that they had learned how to heal some common illnesses. The bishop himself brought blue dye seeds from Dutch Guiana that developed well.

Yet few villagers were happy with the *Semanario*. The majority pointed out that the seeds offered by the *Semanario* were scarce, old, and often did not germinate. In particular, the articles were considered tedious, and the peasants, who according to the bishop were not used to listening to long technical lectures, were bored by the communal readings. Moreover, some peasants got angry when they came to the clergy house. They expected to hear how to cultivate something that they were interested in; however, as soon as the priest had read two or three paragraphs on the subject of interest, the article ended, to be followed on another topic that did not interest them at all (the *Semanario* published articles in installments). As nobody knew when the article of interest would be continued, most of the time peasants did not come back and the priest felt he had wasted his time. Even worse, often neither the peasants nor the priest could understand the texts. What is more, looking into the parish books revealed to the bishop another crucial problem that undermined the rationale of the *Semanario*. Namely, the tithe was well established for crops such as wheat, but not for the new ones. Whenever some peasants managed to grow a new seed, the priest asked for his share. If the peasant refused to pay, as was often the case, the priest discouraged him from cultivating it—"with the excuse that the weather does not favor it," as the bishop reported. In 1805, almost in bankruptcy, the journal was forced to move to the botanical garden. Melón resigned and the *Semanario* became the official journal of the botanical garden. From then on, until its final closure in 1808, it published articles mostly by botanists and chemists, to the exclusion of lay people.

The *Semanario* was deeply at odds with peasants' harsh realities and social indifference. The unequal distribution of lands, the tithe, and the unresponsiveness of higher classes are some of the factors that undermined the pedagogical project and any intention of experimentation that the *Semanario* attempted to bring about. Although it continued to be published after the bishop's report, the government decided not to make subscriptions mandatory, merely reminding the bishoprics of their moral duty to encourage their rectors to buy and use the journal. Some of the bishops wrote back explaining their views. In their answers, several tropes stood out—in particular the "obstinacy" of peasants and their illiteracy, which made them "suspicious of all that they could not see." Furthermore, as the Bishop of Plasencia put it, peasants did not have the means to invest in trials and experiments and "the rich don't care."

The reformers knew this, which is one reason they sought to make alliances with the landed gentry, particularly with landed gentlewomen.

Gardens and Imagined Countries

It was the secretary of the Royal Academy of the Belles Letters, Antonio Ponz, who pompously announced that a vogue for country houses among gentlewomen would benefit society at large: "May God grant that for the benefit of their fatherland the leading gentlewomen [*grandes Señoras*] bring about that which has not been conceded to men."[111] Ponz had traveled extensively through Spain, critically noting barren landscapes and abandoned farms. In his opinion, the remedy lay in influential people investing in agriculture and in particular in gentlewomen who would move others to do the same. He praised the Duchess of Arcos because she transformed her uncultivated land near Madrid into a lush garden of fruit trees, olives, and vineyards and set up a seed nursery and medicinal herb garden for the poor.[112] He highly praised the gardens of the Duchess of Benavente (also known as the Duchess of Osuna),[113] five miles from the Puerta del Sol of Madrid: "The vegetable garden and the plantations that this Lady [Duchess Benavente] had ordered to be built at the site of La Alameda, and the expansion that she thinks of giving it to making it more delicious, useful, and lush, would be in a few years one of the more delightful sites that could be seen."[114]

In fact, El Capricho (The Caprice), as the Duchess named her property, would soon be the chicest country house of Spain. According to contemporaries, El Capricho could easily compete with the royal gardens in Aranjuez. After buying this fertile 150 hectares with a two-floor mansion in 1787, the Duchess of Benavente kept enlarging and improving the property all throughout her life.[115] Some sources suggest that part of El Capricho was devoted to agricultural experiments and that even an agricultural school was set up there. In addition to Ponz, Antonio Hernández de Larrea—the fellow of the Aragón Economic Society who actively defended the inclusion of women in the Madrid Economic Society (see chapter 1)—mentioned the agronomic experiments and a rural school that "gives instruction in the science of the countryside that is more worthy and significant than others instituted only for shouting and disputing without advantage for the republic."[116] Certainly, El Capricho held an impressive number of fruit trees: 1680 pear trees, 536 apple trees, 485 plum trees, 328 apricot trees, 810 grapevines, and ornamental species.[117] Moreover, the gardens were initially designed by Pablo Bouteleou, one of the most highly trained gardeners of Spain, who was acquainted with agronomical trials.[118] Belonging to a family of French gardeners famous in Spain, he was sent to Paris in 1764 to study with the gardeners of the French king, particularly with Richard in Choisy-le-Roi.[119] He then spent a year in Holland and some months in England. He would later collaborate in the Jardín del Príncipe in Aranjuez, in which several botanical projects were tried out.[120] El Capricho also had tree nurseries and a greenhouse, which might have been used for acclimating exotic species. The duchess also bought soil and seeds from the botanical gardens in Paris and Holland.

According to Tom Williamson, "eighteenth-century parks and gardens fulfilled many different roles and expressed many aspects of the lives of their owners."[121] For instance, gardens could express ideals about the role of private persons in contributing to national wealth.[122] In the following, I would like to explore the creative ways in which El Capricho and other noblewomen's gardens may have contributed to social imaginaries of the time.[123] The gardens occupied an interesting status between the private and the public. Although fashioned as an intimate retreat from the court, they also admitted visitors by appointment and regularly hosted parties that competed with the royal ones. El Capricho not only showed off the delicate taste of its feminine owner but also staged an ideal landscape, a utopian country in which women had a crucial role.

El Capricho was designed in the so-called Anglo-Chinese or picturesque style. Arguably based on the descriptions given by the architect Richard Chambers in his *Designs of Chinese Buildings* (1757), this was a supposedly English invention that stood in stark contrast to the grand geometrical French Baroque garden of the past epochs.[124] In the 1780s this style was still a novelty in Spain. Anglo-Chinese gardens had serpentine tracks, irregular topographies, and babbling brooks, but what was most peculiar was the "follies": theatrical settings placed strategically here and there to spark the imagination. Smells, colors, sounds, even the placement of the morning or evening light were all carefully considered to create unforgettable sensual experiences. Although catalogues offered models for these "follies"—the perfect eremite house, the perfect roof for a Chinese pavilion—it was the owner who ultimately created the garden's character. In the case of El

Capricho, the duchess's guests might unexpectedly bump into a tiny chapel with a real hermit, a two-floor cottage in the Petit Trianon style, a navigable channel leading by gondola from a Chinese quay to a dancing pavilion, a hydraulic machine, a column with Saturn eating his son at the top, and a neoclassical apiary housing eighty beehives that could be seen through its glass walls. These sensual experiences were nevertheless laden with moral and political significance. The two-floor cottage, for instance, was customized with a figure of an old woman spinning, so it might symbolize the ideal peasant house, in accordance with ideas of the time. In order to stop rural migration toward urban textile production, Campomanes and his allies proposed to outsource this work to rural women (recall the letters of the *Semanario* in the first part of this chapter). Moreover, the duchess herself promoted spinning among Madrid's poor women, as we saw in chapter 1. The cottage, the fruit trees, the canals and the brooks, the silkworm breeding ground, the agronomic trials, and the rural school—all might invite visions of a prosperous country.[125]

However, in addition to representing an ideal prosperous landscape, the duchess's gardens might have also inspired new female performative roles—in this toing and froing between literary and real worlds, each always inspiring and reinforcing the other. A promenade through El Capricho afforded upper-class female visitors the opportunity not only to play at farming but also to reenact the role of the woman improver, such as the Countess of Jonval in the *Spectacle de la nature*, whose country house was the setting for her attentive observation of nature, as we saw in chapter 1.[126] Consider, for example, the apiary. Transparent beehives and other instruments of observing natural life were widely used by literary and real characters alike for admiring the "wonders of nature." We find evidence of this in the engraving included in the best seller by René-Antoine Ferchault de Réaumur, *Mémoires pour servir à l'histoire des insectes* (Memoirs to contribute to the history of insects [1734–1742]), which shows men and women of polite society observing the bees through a transparent beehive (figure 5.3). Sitting in the apiary of El Capricho, female observers of the bees' hustle and bustle might have recalled and perhaps reenacted not only the Countess of Jonval's study of insects but also her enlightened conversation—always brightened with philosophical and religious reflections—with the other guests in the room. Yet the beehives in El Capricho were placed in an inspiring setting: to observe the bees, visitors sat in a luxurious hall decorated with Italian marble floors and

MÉMOIRES

POUR SERVIR A L'HISTOIRE DES INSECTES.

PREMIER MÉMOIRE.

HISTOIRE DES TIPULES.

L'HISTOIRE des Mouches appellées *Tipules*, auroit été placée dans le dernier volume, s'il eût été poſſible de l'y faire entrer, ſans le rendre d'une groſſeur incommode; il finit par l'hiſtoire des couſins, à la ſuite de laquelle celle des Tipules devoit naturellement ſe trouver; mais au moins cette derniere ne ſera ſéparée de l'autre par aucun

Tome V. A

Figure 5.3. A transparent beehive. In René-Antoine Ferchault de Réaumur, *Mémoires pour servir à l'histoire des insectes* (1734–1742). Vol. 4, p. 53. Max Planck Institute for the History of Science, Berlin.

columns with Corinthian golden capitals. One can only speculate whether the intention of the Duchess of Benavente was to invoke the lesson of Mandeville's *Fable of Bees*: private vices such as desire for material goods could drive the wealth of society and bring public benefit. Albeit in paradoxical ways, El Capricho at once fostered and mirrored the kind of narratives that we discussed at the beginning of this chapter: aristocratic women engaged in improving their families, their states, and their fatherland.

Conclusion

This book has discussed the participation of high-ranking Spanish women in the bustling improvement movement of the late eighteenth century. Improvers, men and women alike, believed that scientific knowledge of a certain kind—which they labeled "useful"—could be a source of happiness for individuals and communities. Knowing how bodies and nature worked, mastering the principles that governed everyday chores, learning the right techniques for work, and prudently administrating social and natural resources would benefit people of all walks of life and, they believed, eventually lead to harmonious couples, loving families, unified communities, and wealthy, populated countries. In order to make and disseminate this useful knowledge, men improvers assembled in societies of "friends of the country," promoted courses for farmers (covering topics such as chemistry, agriculture, and political economy), translated manuals on all kinds of practical arts, experimented with all facets of husbandry and industries, and published extensively on every subject that they considered useful, from children's education to how to bake bread. They thought up initiatives to educate the population on a large scale, for instance, through the extended networks of parish priests. Despite an undoubtedly top-down approach, men improvers sought the collaboration of those who had less influence and importance in the realm of scholarship, either because of their class or gender—in particular, country people and high-ranking women. The experiences of these first collaborations were crucial for collecting locally situated knowledge. The motives these men had for recruiting elite women were varied, ranging from recognizing their moral and social influence to seeking their material help, profiting from their social networks, and benefiting from their expertise in particular areas. In Spain in the 1780s, debates on the intellectual equality of men and women had reached their height. Moreover, the idea that women's behavior impacted both the trade relations and the emotional welfare of families was widely shared; thus, gaining women's endorsement of reformist politics seemed to some to be a wise strategy. In any case, aristocratic women seized their opportunity.

These were politically and philosophically unsettled times in Spain and elsewhere—just before the French Revolution and the Napoleonic Wars—in which the role of aristocracy was already being fiercely questioned in some quarters. One evident strategy of the aristocracy to secure their privileges was to make themselves of service to the community. At the same time, the professionalization of science still lay far in the future, and intellectual elites needed patronage, contacts, and legitimation, which high-ranking women could provide. These "enabling conditions," in the words of Lisbet Koerner, made it possible for a group of learned, powerful aristocratic women from Madrid to organize the female branch of the capital's Economic Society, while from their households, learned women all around Spain tested and annotated recipes, translated manuals, and sent the achievements they considered of public interest to journals and societies.[1] Sharing the contemporaneous ethos of public utility, these women skillfully used contemporaneous gender discourses and the power conferred by their rank gave in order to claim their expertise in certain spaces and areas of knowledge, such as girls' education and the rearing of children, textiles, foundling houses, and domestic economy. They put forward, or better said, they dynamically embodied and created, a new persona—the woman improver—in which erudition and learning was put into service of the society.

Hence it was several historical contingencies—some operating at the European scale, others on the scale of the nation, and even on the more local scale (the court of Madrid)—that permitted a generation of educated women to enter into male domains and wield some kinds of political power. In other words, the existence of the Junta de Damas and other women improvers suggests ways in which women participated publicly in the exciting climate of making and circulating knowledge useful for the community.

This book does not argue that this was an easy way for women to intervene in contemporaneous science. On the contrary, I have shown that women had to skillfully navigate the gender conventions of the time. They had to be cautious when discussing with doctors, when publishing, or when sending their achievements to journals and male societies—otherwise they risked their reputation. Moreover, they often needed powerful allies, either in politics or in scholarly arenas (the Junta, for instance, sought the support of the queen; other women improvers sought the support of fathers, brothers, and other male kin, as we find with María Antonia Gutiérrez Bueno, a woman who carried out experiments in chemistry with the backing of her father, who was a royal apothecary). These women needed to package their activities with powerful rhetoric (carefully orchestrated public performances, touching discourses that exalted their motherly sensibility) and to show "objective" measurements of success, such as mortality rates.

What this book does set out to accomplish, though, is to complicate once again the narrative of scientific exclusion based in gender, showing instead the complicated negotiations and alliances between two groups that needed each other: men improvers and high-ranking women. It also complicates the ways in which we understand what it meant to conduct science in the late eighteenth century, and how practices usually not considered within the spectrum of "scientific" can nonetheless produce scientific knowledge. It shows, for instance, that we have to look more attentively to bureaucratic practices, an issue raised recently by Sebastian Felden and Christine von Oertzen, who advocate "not splitting up areas of expertise."[2] One of the aims of this book is to situate some activities of women hitherto not considered as related to knowledge production, for example in charity, in this context of the scientific improvement movement.

This book also offers historical context for looking at present issues in fresh ways. To a certain extent, we could say that here we are dealing with an early version of citizens' science and communicative practices aiming to remediate the "scientific gap."[3] How to teach girls science and how to engage mothers and lay people are issues that were already raised many years ago by the actors described here. These figures imagined their readers, discussing ideal pedagogical methods and ways to make their materials more attractive and useful. The editors of the *Semanario de Agricultura* (chapter

5), for instance, intended to engage the Spanish rural population in experimenting and collecting local facts and learning together. They dreamed of having informants all across the country; they imagined peaceful Sunday evenings at parish rectories in which countrymen and rectors would come together to discuss the reports of other practitioners and design new trials. Although they succeed in some cases, the reader they had in mind was at odds with the harsh reality of the Spanish countryside: its rampant poverty, unequal distribution of lands, and lack of infrastructure. These are lessons that we could perhaps adopt when questioning why educational policies fail or succeed.

To be sure, this book does not deal with a large population of women. The women whose activities the book unveils were not "an army of volunteers," to cite Sally Shuttleworth's description of how lay people in nineteenth-century Britain participated in the "explosion of print."[14] Neither was Spain experiencing such an explosion—quite the contrary, as fears of contagion from the French Revolution prompted the prime minister to close all journals in 1780s, with the exception of two official publications. None of the women discussed in this book made discoveries that we celebrate today. Nevertheless, highlighting that women improvers were needed and that they did engage in collective enterprises for making and circulating useful knowledge might help us reimagine the future of scientific communities and offer, to the new generations of scientists who still wonder why there are so few women in science, other narratives about this past.

NOTES

Introduction: Women and the Enlightened Campaigns for Improvement

1. The argument was that the ideal place for women to develop all their capacities (moral, emotional, intellectual) was the domestic space due to the particular nature of their bodies. This was of course not the only view; see for instance a good introduction on the unending debates on women's nature and their social roles in Karen O'Brien, "Sexual Distinctions and Prescriptions," in *Women, Gender and the Enlightenment*, ed. Sara Knott and Barbara Taylor (New York: Palgrave Macmillan, 2005), 3–7. For a longer view from the history of science perspective, see Lorraine Daston, "The Naturalized Female Intellect," *Science in Context* 5, no. 2 (Autumn 1992): 209–35, and Lorraine Daston and Gianna Pomata, "The Faces of Nature: 'Visibility and Authority,'" in *The Faces of Nature in Enlightenment Europe*, ed. Lorraine Daston and Gianna Pomata (Berlin: Berliner Wissenschafts-verlag, 2003), 1–16.

2. This first meeting was attended by the Duchess of Benavente, the Countess of Montijo, the Countess of Santa Eufemia, the Marquise of Villalopez, the Marquise of Torrecilla, the Marquise of Ayerve, the Marquise of Palacios, the Countess of Benalua, Doña Maria Rosario Cepeda, and Doña Teresa Losada. Admitted to the Junta, but unable to attend, were the Countess of Fernan Nuñez, the Duchess of Almodovar, the Countess of del Carpio, Doña Felipa de la Rosa, Doña Mariana de Pontejos, and Doña Maria Isidra Quintina Guzmán y de la Cerda. See *Diario Curioso, Erudito, y Comercial*, no. 467, October 10, 1787, 410–12.

3. On the eighteenth-century meaning of "economy" (spelled "oeconomy" in English and German speaking countries), see the groundbreaking work of Lissa Roberts and Simon Werrett: Lissa Roberts, "Practicing Oeconomy during the Second Half of the Long Eighteenth Century: An Introduction," *History and Technology* 30, no. 3 (2014): 133–48 and Simon Werrett, *Thrifty Science: Making the Most of Materials in the History of Experiments* (Chicago: Chicago University Press, 2019). See also: Lissa Roberts and Simon Werrett, "Introduction: A More Intimate Acquaintance," in *Compound Histories: Materials, Governance and Production, 1760–1840*, ed. Lissa Roberts and Simon Werrett (Leiden: Brill, 2017), 1–32; Simon Werrett, "Household Economy and Chemical Inquiry," in Roberts and Werrett, *Compound Histories*, 35–56; Emma Rothschild, *Economic Sentiments: Adam Smith, Condorcet, and the Enlightenment* (Cambridge, MA: Harvard University Press, 2001); Margaret Schabas and Neil Di Marchi, "Introduction to Oeconomies in the Age of Newton," in *Oeconomies in the Age of Newton*, ed. Schabas and Di Marchi, *Annual Supplement to History of Political Economy* 35 (2003): 1–13. Also Elaine Leong, *Recipes and Everyday Knowledge: Medicine, Science and the Household in Early Modern England* (Chicago: University of Chicago Press, 2018). And Esteban de Terreros, *Diccionario castellano con las voces de ciencias y artes y sus correspondientes en las tres lenguas francesa, latina e italiana* (Madrid: Viuda de Ibarra, 1786–1793), s.v. "Economía," defined as "Arte de administrar bien, conducta y gobierno prudente. Dícese no sólo acerca de los bienes, sino de la razón, ciencia etc.: 'Dios hizo con economía todas las cosas'."

4. For a survey see Koen Stapelbroek and Jani Marjane, eds., *The Rise of Economic Societies in the Eighteenth Century: Patriotic Reform in Europe and North America* (Basingstoke: Palgrave Macmillan, 2012). For particular regions, see: Kenneth Hudson, *Patriotism with Profit: British Agricultural Societies in the Eighteenth and Nineteenth Centuries* (London: Hugh Evelyn, 1972); Nicholas Goddard, "Agricultural Literature and Societies," in *The Agrarian History of England and Wales 1750–1850*, ed. G. E. Mingay (Cambridge, UK: Cambridge University Press, 1989), 6:361–83; Henry E. Lowood, *Patriotism, Profit, and the Promotion of Science in the German Enlightenment: The Economic and Scientific Societies 1760–1815* (New York: Garland, 1991). For Spain: Jesús Astigarraga, "Economic Societies and the Politicisation of the Spanish Enlightenment," in *The Spanish Enlightenment Revisited*, ed. Jesús Astigarraga (Oxford: Voltaire Foundation, 2015), 63–81; Gabriel B. Paquette, "State-Civil Society Cooperation and Conflict in the Spanish Empire: The Intellectual and Political Activities of the Ultramarine Consulados and Economic Societies, c. 1780–1810," *Journal of Latin American Studies* 39, no. 2 (May 2007): 263–98. See also: Vicente A. Llombart Rosa and Jesús Astigarraga, "Las primeras 'antorchas de la economía': Las sociedades económicas de amigos del país en el siglo XVIII," in *Economía y economistas españoles*, ed. Enrique Fuentes Quintana (Barcelona: Galaxia Gutenberg, 2000), 3:677–707; Luis Miguel Enciso Recio, *Las sociedades económicas en el siglo de las luces* (Madrid: Real Academia de la Historia, 2010).

5. A survey on the changing meanings of useful knowledge may be found in Karel Davids, "Gatekeeping: Who Defined 'Useful Knowledge' in Early Modern Times?" In *History of Technology*, ed. Ian Inkster (London: Bloomsbury Academic, 2012), 31:69–88. See also Larry Stewart and Kelly Whitmer, "Expectations and Utility in Eighteenth-Century Knowledge Economies," in "Expectations and Utility in Eighteenth-Century Knowledge Economies," ed. Stewart and Whitmer, special issue, *Notes and Records: The Royal Society Journal of the History of Science* 72, no. 2 (June 2018): 111–17; in the same issue, Denise Phillips, "Experimentation in the Agricultural Enlightenment: Place, Profit and Norms of Knowledge-Making in Eighteenth-Century Germany," 159–72; Peter Jones, *Agricultural Enlightenment: Knowledge, Technology and Nature, 1750–1840* (Oxford: Oxford University Press, 2016).

6. Roberts, "Practicing Oeconomy." See in the same volume of *History and Technology* Joppe van Driel, "Ashes to Ashes: The Stewardship of Waste and Oeconomic Cycles of Agricultural and Industrial Improvement, 1750–1800," 177–206 and Elena Serrano, "Making *Oeconomic* People: The Spanish *Magazine of Agriculture and Arts for Parish Rectors* (1797–1808)," *History and Technology* 30, no. 3 (2014): 149–76. See also: Lissa Roberts, "Instruments of Science and Citizenship: Science Education for Dutch Orphans during the Late 18th Century," *Science and Education* 21, no. 2 (February 2012): 157–77 and idem., "Going Dutch: Situating Science in the Dutch Enlightenment," in *The Sciences in Enlightened Europe*, ed. William Clark, Jan Golinski, and Simon Schaffer (Chicago: University of Chicago Press, 1999), 350–88. On working houses as an enlightened project, see: Anna Maerker, "Political Order and the Ambivalence of Expertise: Count Rumford and Welfare Reform in Late Eighteenth-Century Munich," in "Expertise and the Early Modern State," ed. Eric H. Ash, special issue, *Osiris* 25, no. 1 (2010): 213–30.

7. Linda Colley, *Britons: Forging the Nation 1707–1837* (New Haven: Yale University Press 2005), 237–73.

8. Colley, *Britons*, 261.

9. I use here the concept of the social imaginary in the practical, future-oriented sense developed by Lissa Roberts. Initially defined as "collectively imagined forms of social life

and social order," Roberts departed from an intellectually oriented conception to include the material practices that would lead to these "forward-looking futures." She proposes the term "imagineering," which was launched by a 1940s marketing campaign, to describe the practices of late eighteenth-century reformers. See Roberts, "Practicing Oeconomy." See also Sheila Jasanoff, "Future Imperfect: Science, Technology, and the Imaginations of Modernity," in *Dreamscapes of Modernity: Sociotechnical Imaginaries and the Fabrication of Power*, ed. Sheila Jasanoff and Sang-Hyun Kim (Chicago: University of Chicago Press, 2015), 1–49.

10. Karen O'Brien, *Women and Enlightenment in Eighteenth-Century Britain* (Cambridge, UK: Cambridge University Press, 2009).

11. This doesn't mean that men were not also responsible of the well-being of the family and its possessions. On the contrary, as many scholars have discussed, the concrete form that the distribution of tasks in the household took often depended on family arrangement. See Leong, *Recipes and Everyday Knowledge*, 1–18; Werrett, *Thrifty Science*; Amanda Vickery, *The Gentleman's Daughter*, esp. the chapter "Prudent Oeconomy"; idem., *Behind Closed Doors: At Home in Georgian England* (New Haven: Yale University Press, 2009); and Karen Harvey, *The Little Republic: Masculinity and Domestic Authority in Eighteenth-Century Britain* (Oxford: Oxford University Press, 2012). For a survey of the duties of women in Spanish households, see Montserrat Carbonell i Esteller, "Trabajo femenino y economías familiars," in *Historia de las mujeres en España y América Latina*, ed. Isabel Morant Deusa (Madrid: Cátedra, 2005), 2:237–62; also in the same volume, Pilar Pérez Cantó, "Las españolas en la vida colonial," 525–54.

12. See for instance Loïc Charles and Christine Théré, "Les femmes économistes: The Place of Women in the Physiocratic Community," in "Enlightened Female Networks: Gendered Ways of Producing Knowledge (1720–1830)," ed. Anna Maerker, Elena Serrano, and Simon Werrett, special issue of *Notes and Records: The Royal Society Journal of the History of Science* (forthcoming); Harriet Olivia Lloyd, "Rulers of Opinion: Women at the Royal Institution of Great Britain, 1799–1812" (PhD diss., University College London, 2019); Dorota Babilas, "From Female Accomplishment to Botanical Science: Mary Delany's 'Paper Mosaicks,'" *Literature Compass* 10, no. 8 (August 2013): 631–42; Francesca Antonelli, "Scrittura, sociabilita e strategie di persuasione: Marie-Anne Paulze-Lavoisier, secrétaire (1758–1836)" (PhD diss., EHESS-Università di Bologna, Paris, 2021).

13. Carla Bittel, Elaine Leong, and Christine von Oertzen, eds., *Working with Paper: Gendered Practices in the History of Knowledge* (Pittsburgh: University of Pittsburgh Press, 2019), 1–14; Sally Gregory Kohlstedt and Helen Longino, "The Women, Gender and Science Question: What Do Research on Women in Science and Research on Gender and Science Have to Do with Each Other?" in "Women, Gender and Science: New Directions," special issue, *Osiris* 12 (1997): 3–15; Nina E. Lerman, Ruth Oldenziel, and Arwen P. Mohun, eds., *Gender and Technology: A Reader* (Baltimore: Johns Hopkins University Press, 2003); Erika Lorraine Milam and Robert A. Nye, "An Introduction to *Scientific Masculinities*," *Osiris* 30, no. 1 (January 2015): 1–14.

14. Peter Clark, *British Clubs and Societies 1580–1800: The Origins of an Associational World* (Oxford: Oxford University Press, 2002).

15. Margaret Jacobs and Dorothée Sturkenboom, "A Women's Scientific Society in the West: The Late Eighteenth-Century Assimilation of Science," *Isis* 94, no. 2 (June 2003): 217–52; Derya Gurses Tarbuck, "Exercises in Women's Intellectual Sociability in the Eighteenth Century: The Fair Intellectual Club," *History of European Ideas* 41, no. 3 (2015): 375–86.

16. Paula Findlen, "A Forgotten Newtonian: Women and Science in the Italian Provinces," in Clark, Golinski, and Schaffer, *Sciences in Enlightened Europe*, 313–49; Rebecca Messbarger, *The Lady Anatomist: The Life and Work of Anna Morandi Manzolini* (Chicago: Chicago University Press, 2010); Marta Cavazza, *Laura Bassi : Donne, genere e scienza nell'Italia del Settecento* (Milan: Editrice Bibliografica, 2020); Massimo Mazzotti, *The World of Maria Gaetana Agnesi, Mathematician of God* (Baltimore: Johns Hopkins University Press, 2007). For a general panorama, see Rebecca Messbarger, *The Century of Women: Representations of Women in Eighteenth-Century Italian Public Discourse* (Toronto: University of Toronto Press, 2002) and Rebecca Marie Messbarger and Paula Findlen, eds., *The Contest for Knowledge: Debates over Women's Learning in Eighteenth-Century Italy* (Chicago: University of Chicago Press, 2005).

17. Josefa Amar's tutors were Antonio Bermejo and Rafael Casalbón (library curator of the Royal Library). Rosario Cepeda was taught by Antonio González Cañaveras, the author of numerous pedagogical treatises (see chapter 4). Also, many aristocrats, such as the Duke of Osuna, received a license from the Inquisition to read banned French philosophers' works.

18. The *Semanario de Agricultura* (1797–1808) published a series of letters on botany and chemistry for ladies. The *Memorial literario* 1 (1804): 27 advertised *Lettres a Mme de C. sur la botanique, et sur quelques objets de Physique, et d'histoire naturelle, suives d'une méthode élémentaire de botanique*, which could be found in the library of Carretas and Carrera de San Jerónimo.

19. José Pardo-Tomás and Àlvar Martínez, "The Ignorance of Midwives: The Role of Cergymen in Spanish Enlightenment Debates on Birth Care," in *Medicine and Religion in Enlightenment Europe*, edited by Ole Peter Grell and Andrew Cunningham (Aldershot: Ashgate, 2007), 49–62; Montserrat Cabré and Teresa Ortiz, eds., *Sanadoras, matronas y médicas en Europa, siglos XII–XX* (Barcelona: Icaria, 2001); Antonio Lafuente and Juan Pimentel, "La construcción de un espacio público para la ciencia: Escrituras y escenarios en la Ilustración Española," in *Historia de la ciencia y de la técnica en la Corona de Castilla*, ed. Luis García Ballester, 4 vols. (Valladolid: Junta de Castilla y León, 2002), 4:113–55; Antonio Lafuente and Nuria Valverde, *Los mundos de la ciencia en la Ilustración española* (Madrid: Fundación Española para la Ciencia y la Tecnología, 2003); Nuria Valverde, *Actos de precisión: Instrumentos científicos, opinión pública y economía moral en la ilustración española* (Madrid: CSIC, 2007). See also Enrique Perdiguero-Gil, "Popularizing Medicine during the Spanish Enlightenment," in *The Popularization of Medicine, 1650–1850*, ed. Roy Porter (London: Routledge, 1992), 160–93.

20. Lorraine Daston, "Afterwards: The Ethos of the Enlightenment," in Clark, Golinski, and Schaffer, *Sciences in Enlightened Europe*, 495–504, here 495.

21. For an overview of the different peoples and sites of science in the early modern era, see Lorraine Daston and Katharine Park, eds., *The Cambridge History of Early Modern Science*, vol. 3 (New York: Cambridge University Press, 2006).

22. Hedley Brooke, "Natural Theology," in *Science and Religion: A Historical Introduction*, ed. Gary Ferngren (Baltimore: John Hopkins University Press, 2002), 163–75.

23. Daston, "Ethos of the Enlightenment," 502.

24. Lorraine Daston and H. Otto Sibum, "Introduction: Scientific Personae and Their Histories," *Science in Context* 16, nos. 1–2 (March 2003): 1–8.

25. Daston and Sibum, "Scientific Personae," 3.

26. Bernard Bovier de Fontenelle, *Entretiens sur la pluralité des mondes* (Paris: Veuve C. Blageart, 1686); idem., *A Discourse of the Plurality of Worlds* (Dublin: Andr. Crook and Sam. Helsham, 1687). A later English translation by William Gardiner carries the title *Conversations on the Plurality of Worlds* (London: Printed for A. Bettesworth and E. Curll, 1715).

27. Paula Findlen, "Becoming a Scientist: Gender and Knowledge in Eighteenth-Century Italy," *Science in Context* 16, nos. 1–2 (March 2003): 59–87; idem., "The Scientist's Body: The Nature of a Woman Philosopher in Enlightenment Italy," in Daston and Pomata, *Faces of Nature*, 211–36. See also: Steven Shapin, "The Image of the Man of Science," in *The Cambridge History of Science*, vol. 4, *Eighteenth-Century Science*, ed. Roy Porter (Cambridge, UK: Cambridge University Press, 2003), 159–83; Sarah Hutton, "The *Persona* of the Woman Philosopher in Eighteenth-Century England: Catharine Macaulay, Mary Hays, and Elizabeth Hamilton," *Intellectual History Review* 18, no. 3 (January 2008): 403–12; Mineke Bosch, "Persona and the Performance of Identity: Parallel Developments in the Biographical Historiography of Science and Gender, and the Related Uses of Self Narrative," *L'homme: Zeitschrift für feministische Geschichtswissenschaft* 24, no. 2 (2013): 11–22; Mary Terrall, "Metaphysics, Mathematics, and the Gendering of Science in Eighteenth-Century France," in Clark, Golinski, and Schaeffer, *Sciences in Enlightened Europe*, 246–71; Isabelle Lémonon Waxin, "La Savante des Lumières française, histoire d'une persona: Pratiques, représentations, espaces et réseaux" (PhD diss., EHESS, 2019). Unfortunately, I was not able to consult the dissertation while preparing this manuscript.

28. Bernard Bovier de Fontenelle, *Conversaciones sobre la pluralidad de los mundos* (Madrid: Imp. de Villalpando, 1796); Teresa González mentioned Fontenelle in the preface to her work *El estado del Cielo para el año de 1778, arreglado al meridiano de Madrid* (Madrid: Manuel Martín [1778?]) (see an analysis of Teresa González's preface in Inmaculada Urzainqui, *"Catalín." De Rita Barrenecha y otras voces de mujeres en el siglo XVIII* [Vitoria-Gasteiz: Santamaría, 2006]). Rita Caveda, *Cartas selectas de una señora a una sobrina suya, entresacadas de una obra inglesa impresa en Filadelfia, y traducidas al español por Doña Rita Caveda y Solares* (Madrid: Garcia y Compañía, 1800). A biography and study of the *Cartas* may be found in Mónica Bolufer et al., ed., *Mujeres y modernización: Estrategias culturales y prácticas sociales (siglos XVIII–XX)* (Madrid: Instituto de la mujer, 2008), 144–58. Carlos LeMaur, *Discurso sobre la astronomía* ó introduccion al conocimiento de los fenómenos astronómicos, sus leyes, su causa y su aplicacion à los usos de la vida civi*l* (Madrid: Francisco Xavier Garcia, 1762). On the significance of LeMaur, see Antoni Malet, "Newton in the Iberian Peninsula," in *The Reception of Isaac Newton in the Europe*, 3 vols., ed. S. Mandelbrote and H. Pulte (London: Continuum, 2019), vol. 1, chapter 6.

29. On a discussion of the epistemology defended by Pluche, see Ann Blair, "Noël-Antoine Pluche as a Jansenist Natural Theologian," *Intellectual History Review* 26, no. 1 (January 2016): 91–99.

30. Nöel-Antoine Pluche, *Le Spectacle de la Nature, ou Entretiens sur les particularités de la Histoire naturelle qui ont paru les plus propres á rendre les jeunes-gens curieux* (Paris: Vve Estienne, 1732–1750), 8 vol.

31. I mean here women considered in general, not in particular cases. Many academics had feminine helpers. See for instance, Mary Terrall, "Masculine Knowledge, the Public Good, and the Scientific Household of Réaumur," *Osiris* 30, no. 1 (January 2015): 182–202. On the masculinization of useful knowledge in the French Academy, see Terrall, "Metaphysics, Mathematics, and the Gendering of Science"; idem., "Gender and Early Modern Science Cluster. Introduction," in "Gender and Modern Science Cluster," ed. Terrall, special issue of *Configurations* 3, no. 2 (Spring 1995): 135–38; in the same volume, Lisbet Koerner, "Women and Utility in Enlightenment Science," 233–55.

32. More details in chapter 1 of this book.

33. Nina E. Lerman, "Categories of Difference, Categories of Power: Bringing Gender and Race to the History of Technology," *Technology and Culture* 51, no. 4 (October 2010): 893–918.

34. Mónica Bolufer Peruga, "The Enlightenment in Spain: Classic and New Historiographical Perspectives," in *The Routledge Companion to the Hispanic Enlightenment*, ed. Elizabeth Franklin Lewis, Mónica Bolufer Peruga, and Catherine M. Jaffe (London: Routledge, 2019), 3–16, here 6; Joaquín Álvarez Barrientos, "Cultura y política entre siglos," in *Se hicieron literatos para ser políticos: Cultura y política en la España de Carlos IV y Fernando VII*, ed. Joaquín Álvarez Barrientos (Madrid: Servicio de Publicaciones de la Universidad de Cádiz, 2004), 11–24; Pedro Ruíz Torres, "Reformismo e ilustración," in *Historia de España*, ed. Josep Fontana y Ramón Villares (Barcelona: Marcial Pons, 2007), 5:610–15.

35. Londa Schiebinger, *The Mind Has No Sex? Women in the Origins of Modern Science* (Cambridge, MA: Harvard University Press, 1989), especially ch. 2.

36. Mónica Bolufer Peruga, "Debate de los sexos y discurso de progreso en la Ilustración Española," in *Modernidad Iberoamericana: Cultura, política y cambio social*, ed. Francisco Colom González (Madrid: Iberoamericana, 2009), 321–50.

37. Bolufer, "Enlightenment in Spain," 3–16.

38. Jean Sarrailh, *La España Ilustrada de la segunda mitad del siglo XVIII* (Mexico City: Fondo de Cultura Económica, 1957); Richard Herr, *The Eighteenth-Century Revolution in Spain* (Princeton, NJ: Princeton University Press, 1958).

39. Emilio La Parra, *La alianza de Godoy con los revolucionarios: España y Francia a fines del siglo XVIII* (Madrid: CSIC, 1992); Francisco Sánchez-Blanco, *El absolutismo y las luces* (Madrid: Marcial Pons, 2002); idem., *La Ilustración en España* (Madrid: Akal Ediciones, 1997); Equipo Madrid, *Carlos III, Madrid y la Ilustración: Contradicciones de un proyecto reformista* (Madrid: Siglo XXI, 1988); Francisco Aguilar Piñal, *La España del absolutismo ilustrado* (Pozuelo de Alarcón: Espasa, 2005); Antonio Mestre, *Apología y crítica de España en el siglo XVIII* (Madrid: Marcial Pons, 2003); Antonio Bonet and Beatriz Blasco, eds., *Fernando VI y Bárbara de Braganza: Un reinado bajo el signo de la paz (1746–1759)* (Madrid: Real Academia de Bellas Artes de San Fernando, 2002).

40. Roy S. Porter and Mikuláš Teich, eds., *The Enlightenment in National Context* (Cambridge, UK: Cambridge University Press, 1981).

41. Paula Demerson, *María Francisca de Sales Portocarreño, Condesa del Montijo: Una figura de la Ilustración* (Madrid: Editora Nacional, 1975); Paloma Fernández-Quintanilla, *La mujer ilustrada en la España del siglo XVIII* (Madrid: Dirección General de Juventud y promoción Sociocultural, 1981); Margarita Ortega, "El siglo XVIII: Introducción," in *Las mujeres de Madrid como agentes de cambio social*, ed. Margarita Ortega (Madrid: Instituto Universitario de Estudios de la Mujer, 1995), 3–55; Bolufer, *Mujeres y modernización*; Joaquín Álvarez Barrientos, "El modelo femenino en la novela española del siglo XVIII," *Hispanic Review* 63, no. 1 (Winter 1995): 1–18; Valentina Fernández Vargas, ed., *El Madrid de las mujeres: Aproximación a una presencia invisible (1561–1833)* (Madrid: Dirección General de la Mujer, 2007); Emilio Palacios, *La mujer y las letras en la España del siglo XVIII* (Madrid: Ediciones del Laberinto, 2002).

42. See for instance: Marta V. Vicente, *Clothing the Spanish Empire: Families and the Calico Trade in the Early Modern Atlantic World* (Basingstoke: Palgrave Macmillan, 2006); idem., "Textual Uncertainties: The Written Legacy of Women Entrepreneurs in Eighteenth-Century Barcelona," in *Women, Texts and Authority in the Early Modern Spanish World*, ed. Marta V. Vicente and Luis R. Corteguera (Aldershot: Ashgate, 2003), 182–92; Carmen Sarasúa, "The Hardest, Most Unpleasant Profession: The Work of Laundresses in Eighteenth-, Nineteenth-

and Twentieth-Century Spain," in *A Social History of Spanish Labor: New Perspectives on Class, Politics and Gender*, ed. José Piqueras and Vicent Sanz-Rozalén (New York: Berghahn Books, 2008), 64–90.

43. Dorinda Outram, *The Enlightenment* (Cambridge, UK: Cambridge University Press, 2013); Jane Rendall, "Gender, Race and the Progress of Civilization: Introduction," in Knott and Taylor, *Women, Gender and Enlightenment*, 70–74, and the contributions in the same volume of Jenny Mander, "No Woman Is an Island: The Female Figure in French Enlightenment Anthropology" (97–116); Silvia Sebastiani, "'Race, Women and Progress in the Scottish Enlightenment" (75–96); and Sylvana Tomaselli, "Civilization, Patriotism and Enlightened Histories of Woman" (117–35).

44. The list is unending. For a sample see: Theresa Ann Smith, *The Emerging Female Citizen: Gender and Enlightenment in Spain* (Berkeley: University of California Press, 2006); Theresa Ann Smith, "Writing Out of the Margins: Women, Translation, and the Spanish Enlightenment," *Journal of Women's History* 15, no. 1 (Spring 2003): 116–43; Mónica Bolufer Peruga, "Conversations from a Distance: Spanish and French Eighteenth-Century Women Writers," in *A Companion to Spanish Women's Studies*, ed. Xon de Ros and Geraldine Hazbun (Woodbridge: Tamesis, 2011), 175–188; Mónica Bolufer Peruga, "Josefa Amar y Borbón (1749–1733): An Intellectual Woman," in *Women, Enlightenment and Catholicism: A Transnational Biographical History*, ed. Ulrich L. Lehner (London: Routledge, 2018), 50–63; Mónica Bolufer, "Las mujeres en la cultura de la Ilustración," in *Ilustración, ciencia y técnica en el siglo XVIII español*, ed. Enrique Martínez Ruiz y Magdalena de Pazzis Pi Corrales (Valencia: Universidad de Valencia, 2008), 209–32; Inmaculada Urzainqui, "Un nuevo instrumento cultural: La prensa periódica," in *La República de las letras en la España del siglo XVIII*, ed. Álvarez Barrientos, François López, and Inmaculada Urzainqui (Madrid: CSIC, 1995), 125–216; idem., "Nuevas propuestas a un público femenino," in *Historia de la edición y la lectura en España*, ed. Victor Infantes, François López, and Jean-François Botrel (Madrid: Fundación Germán Sánchez Ruipérez, 2003), 481–98; Maria Victoria López Cordón, "Traducciones y traductoras en la España de finales del siglo XVIII," in *Entre la marginación y el desarrollo: Mujeres y hombres en la historia. Homenaje a María Carmen García Nieto*, ed. Cristina Segura y Gloria Nielfa (Madrid: Ediciones del Orto, 1996), 89–112.

45. Sara Knott and Barbara Taylor, eds., *Women, Gender and Enlightenment* (London: Palgrave Macmillan, 2005).

46. Catherine M. Jaffe and Elizabeth Franklin Lewis, eds., *Eve's Enlightenment: Women's Experiences in Spain and Spanish America, 1726–1839* (Baton Rouge: Louisiana State University Press, 2009), 15.

47. Franklin Lewis, Bolufer Peruga, and Jaffe, *Routledge Companion*.

48. Clorinda Donato and Ricardo López, *Enlightenment Spain and the "Encyclopédie Méthodique"* (Oxford: Oxford University Studies in the Enlightenment, 2015); Mestre, *Apología y crítica*, 15–46; Agustí Nieto-Galán, "The History of Science in Spain: A Critical Overview," *Nuncius* 23, no. 2 (January 2008): 211–36; Victor Navarro and William Eamon, eds., *Beyond the Black Legend: Spain and the Scientific Revolution* (Valencia: Universitat de València-CSIC, 2007).

49. Kostas Gavroglu et al., "Science and Technology in the European Periphery: Some Historiographical Reflections," *History of Science* 46, no. 2 (June 2008): 154–75, 154. A synthesis of the arguments may be found in the pathbreaking article: James Secord, "Knowledge in Transit," *Isis* 95, no. 4 (December 2004): 654–72.

50. To quote just a few: Jorge Cañizares-Esguerra, "Iberian Science in the Renaissance: Ignored How Much Longer?" *Perspectives on Science* 12, no. 1 (Spring 2004): 86–124; Antonio Barrera-Osorio, *Experiencing Nature: The Spanish American Empire and the Early Scientific Revolution* (Austin: University of Texas Press, 2006).

51. Juan Pimentel and José Pardo-Tomás, "And Yet, We Were Modern: The Paradoxes of Iberian Science after the Grand Narratives," *History of Science* 55, no. 2 (June 2017): 133–47.

52. Lissa Roberts, "Situating Science in Global History: Local Exchanges and Networks of Circulation," *Itinerario* 33, no. 1 (March 2009): 9–30. See also: idem., "Exploring Global History through the Lens of History of Chemistry: Materials, Identities and Governance," *History of Science* 54, no. 4 (December 2016): 335–61.

53. Charles Rollin, *De la manière d'enseigner et d'étudier les Belles Lettres: Par rapport à l'esprit & au cœur* (Paris: Chez Jacques Estienne, 1726–1728), popularly known as *Tratté des ettudes*. It was translated as: Charles Rollin, *Modo de enseñar, y estudiar las Bellas Letras . . . Trad. . . . por Dª María Cathalina de Caso* (Madrid: Impr. del Mercurio, por Joseph de Orga, 1755).

54. Elena Serrano, "Spreading the Revolution: Guyton's Fumigating Machine in Spain. Politics, Technology, and Material Culture (1796–1808)," in Roberts and Werrett, *Compound Histories*, 106–30.

55. Mónica Bolufer, "Transformaciones culturales: Luces y sombras," in Deusa, *Historia de las mujeres*, 479–510.

56. Sophia Brokmann, *The Science of Useful Nature in Central America: Landscapes, Networks and Practical Enlightenment, 1784–1838* (Cambridge, UK: Cambridge University Press, 2020).

57. On feminist debates at the time in Spain, see Mónica Bolufer Peruga, "'Neither Male, Nor Female': Rational Equality in the Early Spanish Enlightenment," in Knott and Taylor, *Women, Gender and Enlightenment*, 389–409. For a broader picture, see: Armel Dubois-Nayt, Marie-Elisabeth Henneau, and Rotraud von Kulessa, *Revisiter la "querelle des femmes": Discours sur l'egalité/inégalité des sexes en Europe, de 1400 aux lendemains de la Révolution* (Saint-Étienne: Publications de l'Université de Saint-Étienne, 2015). In the same publication: Mónica Bolufer and Montserrat Cabré, "La Querelle des femmes en Espagne: Bilan sur l'histoire d'un débat," 31–45.

58. This was for instance the case of the Countess of Montijo and the Countess of Trullas. See Demerson, *María Francisca de Sales Portocarreño, Condesa del Montijo.*

59. Simon Schaffer, "Enlightenment Brought Down to Earth," *History of Science* 41, no. 3 (September 2003): 257–68. The metaphor is inspired by Ann B. Shteir, *Cultivating Women, Cultivating Science: Flora's Daughters and Botany in England, 1760 to 1860* (Baltimore: Johns Hopkins University Press, 1996).

Chapter 1. Enter the Junta de Damas

1. The title in Spanish was "Maestra y Doctora en Filosofía y Letras Humanas." See *Carolo Tertio regi, catholico, semper, augusto pio . . . obtinenda in philosophia, et humanioribus litteris doctorali laurea in maximo universitatis complutensis theatre* (Madrid: Joachim Ibarra, [1785?]); Antonio Delgado, *Por el magisterio en artes liberales que se confirió en la Real Universidad de Alcalá en 6 de junio de 1785 a la ilustrísima señora doña María Isidra de Guzmán y de la Cerda . . .* (Madrid: Antonio Delgado, 1785).

2. For an analysis of one revision of the *Puellae Doctae* from previous centuries, see Iris M. Zavala, ed., *Breve historia feminista de la literatura española* (Barcelona: Anthropos Editorial, 1997). See also María Jesús Vázquez Madruga, *María Isidra Quintina de Guzmán y la Cerda: La doctora de Alcalá* (Madrid: Centro Asesor de la Mujer, 1999).

3. *Mercurio de España*, June 1785, 187–96; see also *Aparato, ceremonias, y solemnidades que se executaron en la entrada que hizo en la Real Universidad de Alcalá la Excelentísima Señora Doña María Isidra de Guzmán* (publishing house unknown; 1785?).

4. On the university festivities accompanying the awarding of titles, see Jose Luis Peset, "La universidad de Alcalá, la mitra y la corona," *Miscelanea Alfonso IX* 43 (2003): 33–43; José Luis Peset, "Enlightenment and Renovation in the Spanish University," in *Universities and Science in the Early Modern Period*, ed. Mordechai Feingold and Victor Navarro-Brotons (Dordrecht: Springer, 2006), 231–39.

5. *Memorial literario instructivo y curioso de la Corte de Madrid* (June 1785), 147–77. The *Memorial* was received in Cadiz, Seville, Granada, Pamplona, Valladolid, Santander, Salamanca, Oviedo, Murcia, and Barcelona. A well-considered journal that enjoyed a wide readership and tackled a comprehensive range of topics, the *Memorial* was among the five most important Spanish journals in the eighteenth century. See Inmaculada Urzainqui, "Un nuevo instrumento cultural: La prensa periódica," in *La República de las letras en la España del siglo XVIII*, ed. Joaquín Álvarez Barrientos, François López, and Inmaculada Urzainqui (Madrid: Consejo Superior de Investigaciones Científicas, 1995), 125–216, here 170. On the history of exceptional women, see Mónica Bolufer Peruga, "Galerías de 'mujeres ilustres' o el sinuoso camino de la excepción a la norma cotidiana (ss. XV–XVIII)," *Hispania: Revista Española de Historia* 60 (2000): 181–224; and the articles in Montserrat Cabré Pairet and Mónica Bolufer Peruga, eds., "La Querella de las Mujeres: Nuevas perspectivas historiográficas," *Arenal* 20 (2013). (Article citation with previous name. Monica Bolufer is now known without the name "Peruga).

6. The names in Spanish were, respectively, Real Academia Española de la Lengua, Real Sociedad Bascongada de los Amigos del País, and Real Sociedad Económica Matritense de Amigos del Pais. Guzmán y de la Cerda was appointed to the Royal Academy of Language six months before her doctorate (December 28, 1784). She was granted admission as a member with full rights to the Madrid Economic Society in February 1786 and to the Basque Society a year before that. On this point, see *Extracto de las Juntas Generales celebradas por la Real Sociedad Bascongada de los Amigos del Pais, en la Villa de Vergara, por Octubre de 1789* (Vitoria: Baltasar de Manteli, 1789), 38. The discourses that she delivered at the Royal Academy of Language and at the Madrid Economic Society were published in the *Memorial literario* in May 1785, 5–13, and March 1786, 57–361, and as books: María Isidra Quintina de Guzmán y La Cerda, *Oración del género eucarístico, que hizo a la Real Academia Española . . . por Socia de dicha Real Academia* (Madrid: Joaquín Ibarra, 1785) and idem., *Oracion del género eucarístico . . . a la Real Sociedad de Amigos del País* (Madrid: Antonio de Sancha, 1786).

7. See an updated discussion on these questions in Mónica Bolufer Peruga, "Gender and the Reasoning Mind: Introduction," in *Women, Gender and Enlightenment*, ed. Sarah Knott and Barbara Taylor (London: Palgrave Macmillan, 2005), 189–94; and Catherine M. Jaffe, "Contesting the Grounds for Feminism in the Hispanic Eighteenth-Century: The Enlightenment and Its Legacy," in *The Routledge Companion to the Hispanic Enlightenment*, ed. Elizabeth M. Franklin Lewis, Mónica Bolufer Peruga, and Catherine M. Jaffe (London: Routledge, 2019), 69–82.

8. Marta Cavazza, "Between Modesty and Spectacle: Women and Science in Eighteenth-Century Italy," in *Italy's Eighteenth Century: Gender and Culture in the Age of the Grand Tour*, ed. Paula Findlen, Wendy Wassyng Roworth, and Catherine M. Sama (Stanford, CA: Stanford University Press, 2009), 275–302.

9. An online copy of the portrait can be found on the Biblioteca Digital Hispánica: http://bdh-rd.bne.es/viewer.vm?id=0000034026&page=1, accessed August 18, 2020.

10. Mónica Bolufer Peruga, "Debate de los sexos y discurso de progreso en la Ilustración Española," in *Modernidad Iberoamericana: Cultura, política y cambio social*, ed. Francisco Colom González (Madrid: Iberoamericana, 2009), 321–50; Mónica Bolufer et al., eds., *Mujeres y modernización: Estrategias culturales y prácticas sociales (siglos XVIII–XX)* (Madrid: Instituto de la Mujer, Colección Estudios, 2008), 46–54; Mónica Bolufer Peruga, "New Inflections of a Long Polemic: The Debate between the Sexes in Enlightenment Spain," in *A New History of Iberian Feminisms*, ed. Silvia Bermudez and Roberta Johnson (Toronto: University of Toronto Press, 2018), 38–49; Emilio Palacios, *La mujer y las letras en la España del siglo XVIII* (Madrid: Ediciones del Laberinto, 2002); Joaquín Álvarez Barrientos, "El modelo femenino en la novela española del siglo XVIII," *Hispanic Review* 63, no. 1 (Winter 1995): 1–18.

11. Antoine-Léonard Thomas, *Essai sur le caractère, les moeurs et l'esprit des femmes dans les différens siècles* (Paris: Moutard, 1772); Alonso Ruiz de Piña, *Historia o pintura del talento, carácter y costumbres de las mujeres en los diferentes siglos* (Madrid: Miguel Escribano, 1773). For a recent discussion on the history of women as an indicator of the stage of a civilization's development, see Jane Rendall, "Gender, Race and the Progress of Civilization: Introduction," in Knott and Taylor, *Women, Gender and Enlightenment*, 70–74, and the contributions of Jenny Mander, "No Woman Is an Island: The Female Figure in French Enlightenment Anthropology" (97–116); Silvia Sebastiani, "'Race,' Women and Progress in the Scottish Enlightenment" (75–96); and Sylvana Tomaselli, "Civilization, Patriotism and Enlightened Histories of Woman" (117–35) in the same volume.

12. *Memorial literario* XXVIII (April 1786), 473: "No hay nación culta que no pueda presentar un crecido número de mujeres estudiosas y aplicadas."

13. Voltaire's essay was translated into English in 1759, Montesquieu's *Persian Letters* in 1722. On the controversy in Spain, see Antonio Mestre, *Apología y crítica de España en el siglo XVIII* (Madrid: Marcial Pons, 2003), 15–46; Clorinda Donato and Ricardo López, *Enlightenment, Spain and the "Encyclopédie Méthodique"* (Oxford: Oxford University Studies in the Enlightenment, 2015); Clorinda Donato and Manuel Romero, "'Todos los progresos que ha hecho el entendimiento humano': Knowledge, Networking and the Encyclopedic Turn in Enlightenment Spain," in Franklin Lewis, Bolufer Peruga, and Jaffe, *Routledge Companion*, 271–85.

14. *Relación de los ejercicios literarios que la señora doña María del Rosario Cepeda y Mayo,* [. . .], *actuó los días 19, 22 y 24 de septiembre del presente año* (Cádiz: Imprenta Manuel Espinosa de los Monteros, 1768). On Rosario Cepeda, see Pedro Álvarez de Miranda, "¿Una niña en la Academia? El caso de Rosario Cepeda y su orgulloso padre," *Boletín de la Real Academia Española* 82 (2002): 39–45. See also Isabel Azcárate, *Una niña regidora honoraria de la ciudad de Cádiz* (Cádiz: Quórum Libros, 2000).

15. *Copia y recolección de los papeles que, en prosa y verso, han dirigido algunos doctos ingenios de esta ciudad, en debido aplauso del desempeño que en sus actos literarios de los días 19, 22 y 24 de el mes próximo pasado, ejecutó la señora Doña María del Rosario Cepeda* (Cádiz: Imprenta Real de Marina, 1768).

16. Francisco Javier Llampillas, *Ensayo histórico-apologético de la literatura española contra las opiniones preocupadas de algunos escritores modernos italianos*, trans. Josefa Amar y Borbón (Zaragoza: Blas Miedes, 1782–1784). Volume 7 was printed with a different title: *Respuesta del señor abate Lampillas a los cargos recopilados por el señor abate Tiraboschi en su carta al señor abate N.N. sobre el ensayo histórico apologético de la literatura española, traducid traducido del italiano*

por D. Josefa Amar y Borbón (Zaragoza: Blas de Miedes, 1786). A discussion of the significance of this work in the context of Italian disputes about the importance of Spanish literature, as well as in the consolidation of the intellectual status of Josefa Amar, may be found in Elizabeth Franklin Lewis, "Women as Public Intellectuals during the Hispanic Enlightenment: The Case of Josefa Amar y Borbón's *Ensayo histórico-apologético de la literature Española*," in Franklin Lewis, Bolufer Peruga, and Jaffe, *Routledge Companion*, 112–25. On the activities of Josefa Amar y Borbón in the Economic Society of Aragón see: Guillermo Pérez Carrión, "Casual Poverty in the Spanish Enlightenment: Josefa Amar y Borbón and the Real Sociedad Económica Aragonesa de Amigos del País," *Dieciocho: Hispanic Enlightenment* 26, no. 2 (2003): 265–94.

17. On the uses of female prodigies for ambitious fathers, see, for instance, Massimo Mazzotti, "Maria Gaetana Agnesi: Mathematics and the Making of the Catholic Enlightenment," *Isis* 92, no. 4 (December 2001): 657–83, and idem., *The World of Maria Gaetana Agnesi, Mathematician of God* (Baltimore: Johns Hopkins University Press, 2007). On Pope Benedict XIV favoring paid positions for female prodigies in the University of Bologna and membership in the Bolognese Academy of the Institute of Sciences to increase the visibility of the city see: Marta Cavazza, *Laura Bassi. Donne, genere e scienza nell'Italia del Settecento* (Milan: Editrice Bibliografica, 2020); Rebecca Messbarger, *The Century of Women: Representations of Women in Eighteenth-Century Italian Public Discourse* (Toronto: University of Toronto Press, 2002).

18. *Examen al que se presentará Doña Pascuala Caro y Sureda* (Valencia: Benito Monfort, 1781).

19. The body of research on the intellectual role of the elite women during the eighteenth century is huge. For a superb general panorama in European contexts, see Anthony J. La Vopa, *The Labor of the Mind: Intellect and Gender in Enlightenment Cultures* (Philadelphia: University of Pennsylvania Press, 2017) and the essays in the monumental Knott and Taylor, *Women, Gender and Enlightenment*.

20. Theresa Anne Smith, *The Emerging Female Citizen: Gender and Enlightenment in Spain* (Berkeley: University of California Press, 2006), 42–50; Valentina Fernández Vargas, ed., *El Madrid de las mujeres: Aproximación a una presencia invisible (1561–1833)* (Madrid: Dirección General de la Mujer, 2007); María del Carmen Iglesias, "La nueva sociabilidad: Mujeres nobles y salones literarios y políticos," in *Nobleza y sociedad en la España moderna*, ed. María del Carmen Iglesias (Oviedo: Nóbel-Fundación Central Hispano, 1997), 2:179–230; Mónica Bolufer Peruga, "Desde la periferia: Mujeres de la Ilustración en 'Province,'" in *La Real Academia de Bellas Artes de San Carlos en la Valencia Ilustrada*, ed. Romà Calle (Valencia: Universitat de Valencia, 2009), 76–82.

21. Mónica Bolufer Peruga, "'Neither Male, Nor Female': Rational Equality in the Early Spanish Enlightenment," in Knott and Taylor, *Women, Gender and Enlightenment*, 389–409.

22. For instance, the copy of the *Relacion de los ejercicios literarios* in the Biblioteca Nacional de España (signature 7/16143[2]) is bound with the book by Juan Antonio González Cañaveras, *Plan de educacion o Exposición de un nuevo método para estudiar las lenguas, geografía, cronología, historia* (Cadiz: Viuda de D. Antonio de Alcántara, 1784).

23. The literature on the debate is vast. An updated analysis of the contributions, especially of Mme de Levancher and Josefa Amar, can be found in Smith, *Emerging Female Citizen*, 87–107. See also Paloma Fernández Quintanilla, *La IX Duquesa de Osuna: Una Ilustrada en la corte de Carlos III* (Madrid: Ediciones 12 Calles, 2017), 275–96. On the European dimension of the debate, see Mónica Bolufer, "Women in Patriotic Societies: A Spanish Debate in a Euro-

pean Context," in *Society, Ladies, and Philanthropy during the Spanish Enlightenment: La Junta de Damas de Honor y Mérito, 1787–1823*, ed. Catherine M. Jaffe and Elisa Martín-Valdepeñas Yagüe (Baton Rouge: Louisiana State University Press, forthcoming spring 2022).

24. The discourses were published in the *Memorial literario* and in *Espíritu de los mejores diarios literarios que se publican en Europa* (Madrid: Manuel González y Antonio Espinosa, 1787–1790). They were further advertised in the *Mercurio histórico-político*, a journal that had subscribers not only in peninsular Spain but also in the Canary Islands, North Africa, Spanish America, and the United States. Olegario Negrín Fajardo has edited the discourses the debate's participants. I will quote from his edition: Olegario Negrín Fajardo, *Ilustración y educación: La Sociedad Económica Matritense* (Madrid: Editora Nacional, 1984).

25. Negrín Fajardo, *Ilustración y educación*, 170: "No se trata de menos que de igualar a las mujeres con los hombres, de darles asiento en sus juntas y de conferir con ellas materias de gravedad, cosa que parece fuera de orden y aún disparatado."

26. Koen Stapelbroek and Jani Marjanen, eds., *The Rise of Economic Societies in the Eighteenth Century: Patriotic Reform in Europe and North America* (Basingstoke: Palgrave Macmillan, 2012). In Spain, the Junta de damas was later followed by other local female branches of economic societies; see María Consolació Calderón España, "Presencia de la mujer en las reales sociedades económicas de amigos del país (1775–1808)," *Foro de Educación* 12 (2010): 185–231.

27. Smith, *Emerging Female Citizen*, 8–12, and Mónica Bolufer Peruga, *Mujeres e Ilustración: La construcción de la feminidad en la Ilustración española* (Valencia: Institució Alfons el Magnánim, 1998), 371–88, here 371. On economic societies as early developments of citizenship in the ancien régime, see Jesús Astigarraga, *A Unifying Enlightenment: Institutions of Political Economy in Eighteenth-Century Spain (1700–1808)* (Leiden: Brill, 2020); Jesús Astigarraga, "Economic Societies and the Politicisation of the Spanish Enlightenment," in *The Spanish Enlightenment Revisited*, ed. Jesús Astigarraga (Oxford: Voltaire Foundation, 2015): 63–81; Gloria Ángeles Franco Rubio, "Captar súbditos y crear ciudadanos, doble objetivo de los '"Amigos del País'" en el siglo XVIII," *Historia Social* 64 (2009): 3–23; idem., "Las Sociedades Económicas de Amigos del País: Un observatorio privilegiado para la práctica política y el nacimiento de la ciudadanía en el Antiguo Régimen," in *Ilustración, ilustraciones*, ed. Jesús Astigarraga, M. Victoria López-Cordón, and José M. Urkia (Donostia: Real Sociedad Bascongada de los Amigos del País-Sociedad Estatal de Conmemoraciones Culturales, 2011), 1:351–68.

28. On the scope of this movement in Europe, see Stapelbroek and Marjane, *Rise of Economic Societies*, 1–25. See also the introduction of this book for a discussion on useful knowledge. On the role of science in these societies in Spain, see Joaquín Fernández Pérez, "La ciencia ilustrada y las sociedades económicas de amigos del país," in *Carlos III y la ciencia de la Ilustración*, ed. Manuel Selles, José Luis Peset, and Antonio Lafuente (Madrid: Alianza Universidad, 1988): 247–45. For other national contexts, see Henry E. Lowood, *Patriotism, Profit and the Promotion of Science in the German Enlightenment: The Economic and Scientific Societies, 1760–1815* (New York: Garland Publishing, 1991).

29. Many studies have offered close analysis of the membership of local Spanish economic societies. See for instance, Paula Demerson and Jorge Demerson, "La Sociedad Económica de Amigos del País de Ciudad Rodrigo," *Cuadernos de historia moderna y contemporánea* 3 (1982): 35–60; José Francisco Forniés Casals, *La Real Sociedad Económica Aragonesa de Amigos del País* (Zaragoza: Caja de Ahorros de la Inmaculada, 2000); Jesús Astigarraga, *Los Ilustrados vascos: Ideas, instituciones y reformas eco-nómicas en España* (Barcelona: Editorial Crítica, 2003); Luis Miguel Enciso Recio, "Presencia y actividad de los burgueses en las Sociedades Económicas,"

Cuadernos de estudios del siglo XVIII 3–4 (1993–1994): 3–60. For an overview of the Spanish societies, see Vicente Llombart Rosa and Jesús Astigarraga Goenaga, "Las primeras 'antorchas de la economía': Las sociedades económicas de amigos del país en el siglo XVIII," *in Economía y Economistas españoles: la Ilustración*, ed. Enrique Fuentes Quintana (Barcelona: Galaxia Gutenberg, 2000), 677–707; Luis Miguel Enciso Recio, *Las sociedades económicas en el siglo de las luces* (Madrid: Real Academia de la Historia, 2010). For an updated discussion of the boundaries between laypeople and experts, see Jeremy Vetter, "Introduction: Lay Participation in the History of Scientific Observation," *Science in Context* 24, no. 2 (June 2011): 127–41. Paola Bertucci problematizes these categories even further, adding the "artist," who can be defined as an "enlightened artisan." See Paola Bertucci, *Artisanal Enlightenment: Science and the Mechanical Arts in Old Regime France* (New Haven: Yale University Press, 2017). A useful introduction to the issues of practical versus theoretical knowledge is Lissa Roberts, "Introduction," in *The Mindful Hand: Inquiry and Invention from the Late Renaissance to Early Industrialization*, ed. Lissa Roberts, Simon Schaffer, and Peter Dear (Amsterdam: KNAW, 2007), 1–10.

30. Mary Terrall has written extensively about the issue of masculinity in France's Académie des Sciences. See, for instance, Mary Terrall, "Masculine Knowledge, the Public Good, and the Scientific Household of Réaumur," *Osiris* 30, no. 1 (January 2015): 182–202 and "Gendered Spaces, Gendered Audiences: Inside and Outside the Paris Academy of Sciences," *Configurations* 3, no. 2 (Spring 1995): 207–32.

31. For a survey of the European women admitted to academies, as well as the European context of the debate see: Bolufer Peruga, "Women in Patriotic Societies. Also Londa Schiebinger, *The Mind Has No Sex? Women in the Origins of Modern Science* (Cambridge, MA: Harvard University Press, 1989), 11–36.

32. Lisbet Koerner, "Women and Utility in Enlightenment Science," *Configurations* 3, no. 2 (Spring 1995): 233–55, here 237. See also: idem., *Linnaeus: Nature and Nation* (Cambridge, MA: Harvard University Press, 1999).

33. On Josefa Amar's strategies to become elected to the Real Sociedad Económica Aragonesa de Amigos del País, see Constance Sullivan, "Josefa Amar y Borbon and the Royal Aragones Economic Society (with documents)," *Dieciocho: Hispanic Enlightenment* 15 (1992): 1–2 and 95–148, and Franklin Lewis, "Women as Public Intellectuals."

34. Demerson and Demerson, "La Sociedad Económica," 44.

35. Koerner, "Women and Utility," 243.

36. See, for instance, Maxine Berg and Elizabeth Eger, eds., *Luxury in the Eighteenth Century: Debates, Desires and Delectable Goods* (Basingstoke: Palgrave, 2002), especially part 4, "The Female Vice? Women and Luxury," 163–204.

37. Lori D. Ginzburg, *Women and the Work of Benevolence: Morality, Politics, and Class in the Nineteenth-Century United States* (New Haven: Yale University Press, 1990); Elizabeth Franklin Lewis, "Actos de caridad: Women's Charitable Work in Eighteenth-Century Spain," *Dieciocho: Hispanic Enlightenment* 31, no. 2 (2008): 269–84; Alisha Rankin, *Panacea's Daughters: Noblewomen as Healers in Early Modern Germany* (Chicago: University of Chicago Press, 2013).

38. Rebecca Haidt, *Women, Work and Clothing in Eighteenth-Century Spain* (Oxford: Voltaire Foundation, 2011); Carmen Sarasúa, "Technical Innovations at the Service of Cheaper Labour in Pre-Industrial Europe: The Enlightened Agenda to Transform the Gender Division of Labor in Silk Manufacturing," *History and Technology* 24, no. 1 (2008): 23–29 and idem., "Una política de empleo antes de la industrialización: Paro, estructura de la ocupación y salarios en la obra

de Campomanes," in *Campomanes y su obra económica,* ed. Francisco Comin and Pablo Martin (Madrid: Instituto de Estudios Fiscales, 2005), 171–91. See also Rebecca Haidt, "The Wife, the Maid, and the Woman in the Street," in *Eve's Enlightenment: Women's Experiences in Spain and Spanish America, 1726–1839,* ed. Catherine Jaffe and Elizabeth Franklin Lewis (Baton Rouge: Louisiana State University Press, 2009), 115–27. The fundamental text in which Campomanes expressed his ideas about women's textile work was Pedro Rodríguez de Campomanes, *Discurso sobre el fomento de la industria popular* (Madrid: Imprenta de Antonio Sancha, 1774).

39. See, for instance, a contemporary Spanish translation of Justi in Antonio Francisco Puig, *Elementos generales de policia, escritos por el señor Juan Henrique Gottlobs de Justi* (Barcelona: Pi Ferrer, 1784), 138: "cuanto más arregladas tiene las costumbres un Pueblo, tiene mayor proporción para contribuir a su felicidad." Xenophonte's *Oeconomy* was widely read and very much quoted in eighteenth-century Spain. It was again translated in 1786 along with *La economía y los medios de aumentar las rentas públicas de Atenas* (Madrid: Benito Cano, 1786). For a discussion of households as the basis of states and the shifting meaning of oeconomy in the long eighteenth century, see Karen Harvey, *The Little Republic: Masculinity and Domestic Authority in Eighteenth-Century Britain* (Oxford: Oxford University Press, 2012), especially chapter 2, and Simon Werrett, *Thrifty Science: Making the Most of Materials in the History of Experiments* (Chicago: University of Chicago Press, 2019), 1–42.

40. Negrín Fajardo, *Ilustración y educación*, 145: "Induce una especie de violencia meterse oficiosamente en los secretos de las familias . . . Por vía de coacción dificultosamente se podría inclinarlas a la honesta ocupación de los varios ramos de la industria."

41. Negrín Fajardo, *Ilustración y educación*, 135: "Cuando las damas de Francia hicieron moda la lectura y estudio de libros curiosos, infundieron el mismo gusto en los hombres sus apasionados; pues el ejemplo que éstos notaban y el deseo de complacerlas los precisaba a enriquecerse de noticias y disertaciones con que presentarse agradables en sus visitas."

42. Negrín Fajardo, *Ilustración y educación*, 138: "adelantamiento, método, práctica, u otra cosa útil."

43. Negrín Fajardo, *Ilustración y educación*, 135: "El abad de Pluche, en su obra *Espectáculo de la Naturaleza,* introduce una dama principal en muchas conversaciones del cultivo de plantas y flores y otras producciones de la Agricultura, en mi concepto no sólo para hacer sus diálogos más gratos a los lectores, sino principalmente para significar de este modo la propiedad o ninguna extrañeza con que puede parecer y presentarse en los estrados y visitas de las damas esta noble aldeana [la Agricultura]."

44. Nöel-Antoine Pluche, *Le Spectacle de la Nature, ou Entretiens sur les particularités de la Histoire naturelle qui ont paru les plus propres á rendre les jeunes-gens curieux* (Paris: Vve Estienne, 1732–1750), 8 vol. The contents of the volumes were "I. Ce qui regarde les animaux et les plantes. 1732"; "II. Ce qui regarde les dehors et l'intérieur de la terre. 1735"; "III. Ce qui regarde les dehors et l'intérieur de la terre. 1735"; "IV. Ce qui regarde le ciel et les liaisons des différentes parties de l'univers avec les besoins de l'homme. 1739"; "V. Ce qui regarde l'homme considéré en lui-même. 1746"; "VI–VII. Ce qui regarde l'homme en société. 1746"; "VIII. 1–2. Ce qui regarde l'homme en société avec Dieu. 1750." The English translation was published in London beginning in 1733.

45. Ann Blair, "Noël-Antoine Pluche as a Jansenist Natural Theologian," *Intellectual History Review* 26, no. 1 (2016): 91–99.

46. Daniel Mornet, "Les enseignements des bibliothèques privées (1750–1780)," *Revue d'Histoire Littéraire de France* 18 (1910): 449–96. See also Blair, "Nöel-Antoine Pluche," 91 and 92.

47. Pluche was a tutor to the youngest son of Lord William Stafford Howard, second Earl of Stafford. On Pluche's biography see Dennis Trinkle, "Nöel-Antoine Pluche's 'Le spectacle de la nature': An Encyclopaedic Best Seller," *Studies on Voltaire and the Eighteenth Century* 358 (1997): 93–134; Françoise Gevrey, Julie Boch, and Jean-Louis Haquette, eds., *Écrire la nature au XVIIIe siècle: Autour de l'abbé Pluche* (Paris: PUPS, Presses de l'Université Paris-Sorbonne, 2006); Benoît De Baere, *Trois introductions à l'Abbé Pluche: Sa vie, son monde, ses livres* (Geneva: Droz, 2001); Robert Loqueneux, "L'abbé Pluche, ou l'accord de la foi et de la raison à l'aube des Lumières," *Sciences et techniques en perspective* 2 (1998): 235–88. See also the biography of Pluche ("Eloge historique de monsieur l'abbé Pluche") that Robert Estienne included in his edition of Antoine-Noel Pluche, *Concorde de la géographie des différens âges* (Paris: Chez les frères Estienne, 1764).

48. Cynthia Koepp, "Curiosity, Science, and Experiential Learning in the Eighteenth Century: Reading the Spectacle de la nature," in *Childhood and Children's Books in Early Modern Europe (1550–1800)*, ed. Andrea Immel and Michael Witmore (New York, 2006), 153–80.

49. Pluche explained this use of the bibliography: see *Le Spectacle*, I, "Preface," xiii.

50. Examples are Filippo Buonani (1638–1725), Johannes Jonston (1603–1675), Ferdinando Marsigli (1658–1730), or Claude Perrault (1613–1688). See Madeleine Pinault-Sørensen, "Les planches du Spectacle de la nature de l'abbé Pluche," in Gevrey, Boch, and Haquette, *Écrire la nature au XVIIIe siècle*, 441–59.

51. Pluche, *Le Spectacle*, vol. 1, "Preface," xv: "Nous trouvons flattés de l'apprendre de nos Semblables: en les entendant on se croit capable de penser."

52. For a discussion on scientific personae, see the introduction to this book.

53. Bernard Bovier de Fontenelle, *Entretiens sur la pluralité des mondes* (Paris: Veuve C. Blageart, 1686); *A Discourse of the Plurality of Worlds* (Dublin: Andr. Crook and Sam. Helsham, 1687). A later English translation by William Gardiner carries the title *Conversations on the Plurality of Worlds* (London: Printed for A. Bettesworth, and E. Curll, 1715).

54. To be sure, scientific literature for women existed before the *Entretiens*, such as René Bary, *La fine philosophie, accomodee a l'intelligence des dames* (Paris: Chez Simeon Piget et chez l'autheur, 1660). However, it was not as successful as Fontenelle's work. See Jeanne Peiffer, "La litterature scientifique pour les femmes au xviii siècle," in *Sexe et genre: De la hierarchie entre les sexes*, ed. Marie-Claude Hurtig, Michele Kail, and Helene Rouch (Paris: Centre National de la Recherche Scientifique, 1991), 138–46; Geoffrey Sutton, *Science for a Polite Society: Gender, Culture, and the Demonstration of Enlightenment* (Boulder, CO: Westview, 1995).

55. For example, in his *The Young Gentlemen and Lady's Philosophy* (1759), Martin also imagined a man instructing a woman on astronomy, but because a lady conversing with a passionate philosopher was considered unacceptable in Britain, he reframed the scene as taking place with an elder brother in a cozy indoor drawing room, with a telescope and a globe at hand (which incidentally were sold in his London shop). See Mary Terrall, "Natural Philosophy for Fashionable Readers," in *Books and the Sciences in History*, ed. Marina Frasca-Spada and Nick Jardine (Cambridge, UK: Cambridge University Press, 2000), 239–254; idem., "Gendered Spaces, Gendered Audiences."

56. See, for instance, Massimo Mazzotti, "Newton for Ladies: Gentility, Gender and Radical Culture," *British Journal for the History of Science* 37, no. 2 (June 2004): 119–46 and Ralph O'Connor, "Reflection on Popular Science in Britain: Gender, Categories and Historians." *Isis* 100, no. 2 (June 2009): 333–45. See also the cases of the chemical and botanical treatises for women published in a rural Spanish journal discussed in chapter 4 of this book. For a wider

overview, see: Agustí Nieto-Galán, *Science in the Public Sphere: A History of Lay Knowledge and Expertise* (London: Routledge, 2016).

57. Paula Findlen, "Translating the New Science: Women and the Circulation of Knowledge in Enlightenment Italy," *Configurations* 3, no. 2 (Spring 1995): 167–206; idem., "The Scientist's Body: The Nature of a Woman Philosopher in Enlightenment Italy," in *The Faces of Nature in Enlightenment Europe*, ed. Gianna Pomata and Lorraine J. Daston (Berlin: Berliner Wiss.-Verl., 2003), 211–36; John Mullin, "Gendered Knowledge, Gendered Minds: Women and Newtonianism, 1690–1760," in *A Question of Identity: Women, Science and Literature*, ed. Marina Benjamin (New Jersey: Rutgers University Press, 1993), 41–58; Nicholas Jardine, James Secord, and Emma C. Spary, eds., *The Cultures of Natural History* (Cambridge, UK: Cambridge University Press, 1996).

58. Paula Findlen, "Becoming a Scientist: Gender and Knowledge in Eighteenth-Century Italy," *Science in Context* 16, nos. 1–2 (March 2003): 59–87, 61; idem., "Women on the Verge of Science: Aristocratic Women and Knowledge in Early Eighteenth-Century Italy," in Knott and Taylor, *Women, Gender and Enlightenment*, 265–87.

59. Terrall, "Masculine Knowledge." See also Paula Findlen, who complicates even more the gendered of knowledge and practitioners in her analysis of Algarotti's *Newtonianism for Ladies* (1737), in which Algarotti included two kinds of women natural philosophers, the women scientists, modeled on the successful Italian luminaries of Bologna, Naples, Venice, and Milan, and the less educated. Findlen, "Becoming a Scientist."

60. For more on the duties of wives and their relation to household economy, see chapter 2 in this book.

61. See the discussion of these two different epistemologies in Bernadette Bensaude-Vincent and Christine Blondel, eds., *Science and Spectacle in the European Enlightenment* (Aldershot: Ashgate, 2008), 8. For a discussion of the relationship between natural philosophy and spectacles, see Simon Werrett, "Watching the Fireworks: Early Modern Observation of Natural and Artificial Spectacles," *Science in Context* 24, no. 2 (June 2011): 167–82. See also Christophe Martin, "Women's Curiosity and Its Double at the Dawn of the Enlightenment," in *Women and Curiosity in Early Modern England and France*, ed. Line Cottegnies, Sandrine Parageau, and John J. Thompson (Leiden: Brill, 2016), 175–96, here 179.

62. See for instance: "But if we desire to fathom the very Depths of Nature, to trace Effects up, to their particular Causes, and comprehend the Curiosity and elastic Play of every secret Spring [a reference to Cartesianism], as well as the minutest Elements that compose them, this is an arduous Attempt." In Pluche, *Spectacle de la nature: or, Nature display'd*, "Preface," viii. In French, see Pluche, *Spectacle*, "Preface," v–vi: "Mais prendre pénétrer le fond même de la Nature, vouloir rappeler les effets à leurs causes spéciales, vouloir comprendre l'artifice & le jeu des ressorts, & les plus petits éléments don ces resots sont composés, c'est une entreprise hardie & d'un succès trop incertain."

63. Pluche, *Spectacle*, 300 (English translation). In French: Pluche, *Spectacle*, 1:542: "Sur toutes les choses créées qui sont sous nos yeux, il ne peut avoir pour nous que l'un de ces trois partis à prendre: L'un serait, de n'en vouloir tout comprendre, le denier seroit d'en vouloir rien connaître; le dernier serait, d'en rechercher, & d'en mettre a profit ce qu'on en peut savoir."

64. Pluche, *Spectacle* (French): vol. 1, "Entr.," xi, 299: "Madame, le mérite des Physiciens, parmi lesquels nous vous comptons à présent, ne consiste pas toujours à deviner des choses difficiles; mais à ouvrir les yeux sur ce les autres n'aperçoivent pas, & qu'ils soulent aux pieds le plus souvent. Rien de plus rare que des gens qui pensent & qui réfléchissent."

65. Elena Serrano, "The *Spectacle de la nature* in Eighteenth-Century Spain: From French Households to Spanish Workshops," *Annals of Science* 69 (2012): 257–82.

66. Esteban de Terreros y Pando, *Diccionario castellano con las voces de ciencias y artes y sus correspondientes en las tres lenguas, francesa, latina é italiana. Su autor el P. Esteban de Terreros y Pando*, 4 vols. (Madrid: Viuda de Ibarra, 1786–1793).

67. Terreros, *Espectáculo*, preface: "Quien no ve, que a V. Mag, y al Espectáculo de la Naturaleza los señala un mismo carácter, que miran a un mismo fin, y que tienen las mismas ideas?"

68. Terreros, *Espectáculo*, preface.

69. Terreros, *Espectáculo*, preface: "Todos pueden saber Lógica . . . todos una Historia, todos una Physica práctica y experimental que levante sus corazones al Criador, y todos, sin discreción de sexos ni edades deben estar instruidos a fondo en su Religión." In English: "All can learn Logic . . . all history, all a practical and experimental physics that might raise their hearts to the Almighty, and all, independent of their sex or age, must be deeply instructed in their religion."

70. Benito J. Feijoo, *Cartas Eruditas y Curiosas*, 5 vols. (Madrid: publishing house unknown, 1742–1760), v, map 23 (1726), 367–91.

71. Juan Sempere y Guarinos, *Reflexiones sobre el Buen Gusto en las ciencias y en las artes. Traducción libre* (Madrid, 1782), 279: "Hasta en los estrados, y entre las Damas llegó a hacerse moda el hablar de la Historia Natural de los animales, de las plantas, y de los minerales; y de los oficios, y fábricas, asuntos enteramente desconocidos fuera de la clase de los artesanos, y de bien pocos facultativos."

72. The *Memorial literario instructivo y curioso* published a summary of the debate in its October 1787 issue, 203–9. Spanish feminist historiography has produced detailed analyses of the debate. Especially interesting are the studies that put the debate into a long-term perspective; see, for instance, Elizabeth Lewis Franklin, "Feijoo, Josefa Amar y Borbon and the Feminist Debate in Eighteenth-Century Spain," *Dieciocho: Hispanic Enlightenment* 12 (1989): 188–203; Smith, *Emerging Female Citizen*, 17–39; and Mónica Bolufer Peruga, "Mujeres y hombres en los espacios del Reformismo Ilustrado: Debates y estrategia," *hMiC: història moderna i contemporània* 1 (2003): 155–70.

73. Francisco Cabarrús, "Discurso," *Memorial literario* 29 (May 1786), 74–85; Melchor Gaspar de Jovellanos, "Discurso en que se prueba que las Señoras deben ser admitidas por Socias," *Memorial literario* 28 (April 1786): 475–88.

74. Josefa Amar y Borbón, "Discurso en defensa del talento de las mujeres, y de su aptitud para el gobierno, y otros cargos en que se emplean los hombres," *Memorial literario* 32 (August 1786): 400–430. An English translation can be found in Josefa Amar y Borbón, *In Defence of Women*, trans. Joanna M. Barker (Cambridge, UK: Salisbury House, 2018), 97–112.

75. *Espíritu de los mejores diarios literarios que se publican en Europa*, no. 73, 675–77; no. 74, 683–85; no. 76, 700–701; no. 77, 708–10. After the Crown's decision, the well-known member of the Madrid Economic Society Ignacio Ayala still presented a new memoire backing Amar y Borbón's ideas (August 30, 1787).

76. Melchor Gaspar de Jovellanos, "Discurso en que se prueba que las Señoras deben ser admitidas por Socias." *Memorial literario* 32 (April 1786): 475–88.

77. Juan Antonio Hernández de Larrea,"Carta a Doña Josefa Amar diciendo su parecer sobre el discurso antecedente," *Memorial literario* 32 (August 1786): 430–38: "No aren, no caben, no sieguen no lleven cargas pesadas según lo practican nuestras Montañesas y serranas, más [las Damas] pueden y deben dirigir y cuidar las cosechas de seda, y sus hilados como hacen

las Emperatrices de la China; pueden cultivar las hortalizas y flores; pueden y deben gobernar la cría de todas las aves; preparar en casa las simientes con las legías que las hacen fecundas, y hay otras mil operaciones rústicas que las pertenecen con las cuales pueden dar impulso a la agricultura."

78. On the cultural context of the debates, see María Victoria López-Cordón Cortezo, *Condición femenina y razón ilustrada: Josefa Amar y Borbón* (Zaragoza: Prensas universitarias de Zaragoza, 2005); idem., "Josefa Amar y Borbón (1749–1833): Los debates ilustrados sobre las mujeres," in *Feminismos: Contribuciones desde la historia*, ed. Ángela En Cenarro and Régine Illion (Zaragoza: Prensas Universitarias de Zaragoza, 2014), 51–80.

79. Borbón, "Discurso en defensa," 400–430. Negrín Fajardo, *Ilustración y educación*, 170: "que las mujeres defiendan su causa, porque el silencio en esta ocasión confirmaría el concepto que se de ellas se tiene de que no se cuidan ni se interesan por asuntos serios."

80. Negrín Fajardo, *Ilustración y educación*, 164–65: "Saben ellas que no pueden aspirar a ningún empleo ni recompensa pública. Que sus ideas no tienen más extensión que las paredes de una casa o de un convento. Si no es esto suficiente para sofocar el mayor talento del mundo, no sé qué otras trabas pueden buscarse."

81. Negrín Fajardo, *Ilustración y educación*, 165. On Amar y Borbón's relationship with Catholicism, see Mónica Bolufer, "Josefa Amar y Borbón (1749–1733): An Intellectual Woman," in *Women, Enlightenment and Catholicism: A Transnational Biographical History*, ed. Ulrich L. Lehner (London: Routledge, 2018), 50–63.

82. Constance A. Sullivan, "Constructing Her Own Tradition: Ideological Selectivity in Josefa Amar y Borbón's Representation of Female Models," in *Recovering Spain's Feminist Tradition*, ed. Lisa Vollendorf (New York: Modern Languages Association of America, 2001), 142–59, here 154.

83. Stéphanie Félicité du Crest de Saint-Aubin, Countess of Genlis (1746–1830).

84. Negrín Fajardo, *Ilustración y educación*, point 34: "Si actualmente ocupa una mujer la Presidencia de las ciencias en una Corte de Europa, que es más que sentarse como individuo en un cuerpo, las materias que trata nunca son tan abstractas; y si en fin se trata de hacerlas amigas del país, . . . lejos de ser perjudicial la admisión las mujeres, puede y debe ser conveniente."

85. Negrín Fajardo, *Ilustración y educación*, "Cotejar entre las obras del extranjero, y nuestras, para ver lo que se puede adelantar, hacer nuevos inventos en las artes, . . . aunque son cosas que piden meditación, y noticias, no son materias tan abstractas, que no las pueda comprender la mujer que tenga talento regular."

86. "El rey entiende que la admisión de socias de mérito y honor, que en juntas regulares y separadas traten de los mejores medios de promover la virtud, la aplicación y la industria en su sexo, será muy conveniente en la Corte . . . , y tratasen unidas los medios de fomentar la buena educación, mejorar las costumbres con sus ejemplos y sus escritos, introducir el amor al trabajo, cortar el lujo que. . . ." In Negrín Fajardo, *Ilustración y educación*, 14; and in *Memorial literario* 39 (October 1787): 204–9; and *Gazeta de Madrid*, October 12, 1787, 688–89.

87. *Diario Curioso, Erudito, y Comercial*, the *Gazeta de Madrid*, the *Mercurio de España*, and the *Memorial Literario* published articles on the first meeting of the Junta. *Diario Curioso, Erudito, y Comercial*, no. 467 (Wednesday, October 10, 1787), 410–12. The ladies that attended this first meeting were: the Duchess of Benavente, Countess of Montijo, Countess of Santa Eufemia, Marquise of Villalópez, Marquise of Torrecilla, Marquise of Ayerve, Marquise of Palacios, Countess of Benalua, Doña Maria Rosario Cepeda, Doña Teresa Losada. Admitted

to the Junta but not able to come to this first meeting were: the Countess of Fernán Nuñez, Duchess of Almodóvar, Countess of del Carpio, Doña Felipa de la Rosa, Doña Mariana de Pontejos and Doña Maria Isidra Quintina Guzmán y de la Cerda.

88. *Gazeta de Madrid*, October 12, 1787, 890.

89. Elisa Martín-Valdepeñas Yagüe, "La reina María Luisa de Parma y la Junta de Damas de la Sociedad Económica Matritense," in *La época de Carlos IV (1788–1808): Actas del IV Congreso Internacional de la Sociedad Española de Estudios del Siglo XVIII*, ed. Elena de Lorenzo Álvarez (Madrid: Ediciones Trea, 2009), 749–62. The Biblioteca Nacional de España kept some of the discourses (*Elogios de la Reina*) addressed to the queen during the annual distribution of prizes to the girls in the four spinning schools that the Junta ran, including those by the Countess of Torrepalma, María Francisca Dávila Carrillo de Albornoz (1794); the Marquise of Sonora, María Josefa Valenzuela (1796); the Marquise of Fuerte-Híjar, María Lorenza de los Ríos y Loyo, (1798); Josefa Díez de la Cortina (1800); and the Countess of Castrocarreño (1801).

90. She was president of the Junta during intermittent periods: 1787–1790, 1801–1811, and 1814–1817.

91. The duchess married Pedro de Alcántara Téllez Girón y Pacheco (1756–1807), the Marquis of Peñafiel, Count of Ureña, Duke of Osuna. See Fernández Quintanilla, *La IX Duquesa de Osuna*; Ignacio Atienza Hernández, *Aristocracia, poder y riqueza en la España Moderna: La Casa de Osuna, siglos XV–XIX* (Madrid: Siglo veintiuno de España, 1987); idem., "Mujeres que mandan: Aristócratas y ciclo vital en el siglo XVIII," in *Historia de las mujeres en España y América Latina*, vol. 7, ed. Isabel Morant Deusa, 437–76 (Madrid: Cátedra, 2006). Also Carmen Muñoz de Figueroa Yebes, *La Condesa-Duquesa de Benavente: Una vida en unas cartas* (Madrid: Espasa-Calpe, 1955).

92. Her secretary Manuel de Ascargorta was a member of the Academy of History, while the family doctor Hilario Torres belonged to the Royal Academy of Medicine. Her children's painting teacher was the celebrated painter Manuel Esteve, and the erudite Diego Clemencín was their tutor. The duke had one of the largest libraries in Spain. For an idea of how much it cost to maintain their lavish lifestyle, see AHN, "Budget for Employees," in Osuna, *Cartas*, 452; "Will" (n.d.): Osuna, ct. 519; "Expenses" (including the tutors of the children): Osuna, ct. 515.

93. Charles de Pougens, *Mémoires et souvenirs de Charles de Pougens, chevalier de plusieurs ordres . . .* (Paris: Librarie de H. Fournier Jeune, 1834), 193.

94. When she died, the French ambassador in Madrid, Maximilien Gérard de Rayneval (1778–1836) said about her: "C'est une parte irréparable pour la Sociéte de Madrid. Son Salon depuis plus d'un demi-siècle était le rendez-vous habituel de tout ce qu'il y avait de plus distingué dans la capitale, Espagnols et étrangers." Arch. Aff. Etr., Paris, Spain, Corr. Pol, 766, 122 r-v.

95. Nicolás de Toruneaux, *Instrucciones cristianas sobre el sacramento del matrimonio* (Barcelona: Bernardo Pla, 1774). See Paula Demerson, *María Francisca de Sales y Portocarrero, condesa del Montijo: Una figura de la Ilustración* (Madrid: Editora Nacional, 1975).

96. Elena Serrano, "Sex and Prisons: Women and Spanish Penitentiary Reform, 1787–1808," in "The Sexes and the Sciences," ed. D. N. Wagner and J. Wharton, special issue, *Journal for Eighteenth-Century Studies* 42, no. 4 (December 2019): 501–17.

97. See for instance, Glenda Sluga and Carolyn James, eds., *Women, Diplomacy and International Politics since 1500* (New York: Routledge, 2016).

98. For instance, the Marquise of Fuerte-Híjar, president of the Junta from 1811–1814, was

married to a member of the Consejo de Castilla and officer of the Theater (subdelegado de teatros); the Countess of Torrepalma y Trullas (who was president of the Junta from 1790 to 1801) was married to a diplomat; the first Duchess of Almodóvar was married to the ambassador in Lisbon and London; the Marquise of Llano, to the ambassador in Vienna.

99. *Estatutos de la Junta de Socias de Honor y Mérito de la Real Sociedad Económica de Madrid* (Madrid: Sancha, 1794), 3–4: "Para ser socia es necesario una buen educación y conducta con instrucción notoria en los objetos del Instituto."

100. *Estatutos de la Junta*, 5: "Estatuto VIII": "Las ausentes ejecutarán los experimentos y comunicarán las noticias que se las encarguen."

101. Francesco Maria Cavazzoni Zanotti, *Compendio de la Filosofia Moral, traducido del Italiano por la Marquesa de Espeja* (Madrid: Joaquín Ibarra, 1785). She was married to Antonio del Águila, Marqués de Espeja, Alférez mayor of Ciudad Rodrigo.

102. Catherine M. Jaffe and Elisa Martín-Valdepeñas Yagüe, *María Lorenza de los Ríos, marquesa de Fuerte-Híjar. Vida y obra de una escritora del Siglo de las Luces* (Madrid: Iberoamericana-Vervuert, 2019).

103. The list of members through time can be found in Smith, *Emerging Female Citizen*, 137–39, table 5. See also Paula de Demerson, "Catálogo de las socias de honor y mérito de la Junta de Damas Matritense," in *Anales del Instituto de Estudios Madrileños* 7 (1971): 269–74.

104. The Marquise of Fuerte-Híjar set up a female branch of the Vallodolid Economic Society. See Catherine M. Jaffe and Elisa Martín-Valdepeñas Yagüe, "Sociabilidad, filantropía y escritura: María Lorenza de los Ríos y Loyo, Marquesa de FuerteHíjar (1761–1821)," in *Mujeres y culturas políticas en España, 1808–1845*, ed. Ana Yetana Laguna (Barcelona: Universitat Autónoma de Barcelona), 87–128; Alberto Acereda, *La Marquesa de Fuerte-Híjar: Una dramaturga de la Ilustración* (Cádiz: Universidad de Cádiz, 2000); idem., "Una figura relegada de la Ilustración: La Marquesa de Fuerte-Híjar y su Elogio a la Reina (1798)," *Cuadernos de Investigación Filológica* 23–24 (1997–1998): 195–212.

105. February 1, 1788. She answered on March 14; see Demerson, *María Francisca de Sales*, 165–68.

106. Archivo Sociedad Ecónomica Matritense (SEM) (1794): "Razón de lo suplido de orden de las señoras y señores comisionados para los experimentos sobre los tintes firmes que ofrecen hacer Manuel Balius? and Carlos Rivera."

107. SEM: "Letter from Pedro Gómez Ibar to the Junta: Libro de Actas de las damas (session of July 18, 1788)"; "Letter from Manuel de Fox to the Junta" (session of September 5, 1788); María Guerrero sent samples of silk and a list of the Catalan silk manufactures (August, 15, 1788); Carlos Guardia sent velvet and cotton samples from Barcelona (October 3, 1788); Juan Nadal presented linen and hemp textile samples (August 8, 1788) and textiles from Yébedo (March 14, 1790); Andrea de Varo in Aguilar de la Frontera sent linen and different cottons (April 30, 1790). See also Demerson, *María Francisca de Sales*, 166–67.

108. Smith, *Emerging Female Citizen*, 173, 308. Along with the Marquise of Canillejas, the Marquise of Fuerte-Híjar tested a machine for silk twisting. Quoted in José Varela de Limia y Menéndez [Vizconde de San Alberto], *Los directores de la Real Sociedad Económica Matritense de Amigos del País y las presidentas de su Junta de Damas de Honor y Mérito* (Madrid: Talleres de "El Eco Franciscano," 1925).

109. Juan Sempere y Guarinos, *Ensayo de una biblioteca española de los mejores escritores del reinado de Carlos III* (Imprenta Real, Madrid: 1785–1789), vol. 5 (1789), 213: "Till the reign of Carlos III there has not been seen in Spain any society of women authorized by the sovereign

except monasteries, congregations, brotherhoods, and other committees devoted only to certain exercises of piety and devotion." ["Hasta el reinado de Carlos III no se ha visto en España ninguna asociación de mujeres autorizada por el soberano, a excepción de los monasterios, congregaciones, cofradías y otras juntas dirigidas únicamente a ciertos ejercicios de piedad y devoción."]

110. On a recent historiographic discussion on the molding of public identity in relation to gender and science, see Mineke Bosch, "Persona and the Performance of Identity: Parallel Developments in the Biographical Historiography of Science and Gender, and the Related Uses of Self Narrative," *L'homme: Zeitschrift für feministische Geschichtswissenschaft* 24, no. 2 (2013): 11–22. Bosch relies on the seminal article of Lorraine Daston and H. Otto Sibum, "Introduction: Scientific Personae and Their Histories," *Science in Context* 16, nos. 1–2 (March 2003): 1–8.

Chapter 2. Putting Order in the Foundling House

1. These were the Marquise of Ariza, the Countess of Trullas, the Countess of Montijo, and Doña Francisca Raón y Mariño. On the situation of the Madrid Foundling House, see Fausto Martínez de la Torre, *Plano de la Villa y Corte de Madrid: En sesenta y quatro láminas, que demuestran otros tantos barrios en que está dividida, con los nombres de todas sus plazuelas y calles* (Madrid: Imprenta D. Joseph Doblado, 1800).

2. Annual reports of the Junta de Honor y Mérito, *Memorias instructivas de los negocios de la Junta de Honor y Mérito,* 8880/17, Archivo Regional de la Comunidad de Madrid (ARM), see especially the report of 1800. These were annual reports about the activities of the Junta to be read at the Madrid Economic Society at the end of the year. Hereafter, I refer to them as the Annual Report. See also Paula Demerson, *María Francisca de Sales y Portocarreño, condesa del Montijo: Una figura de la Ilustración* (Madrid: Editora Nacional, 1975), 215–44; Joan Sherwood, *Poverty in Eighteenth-Century Spain: The Women and Children of the Inclusa* (Toronto: Toronto University Press, 1988), 9–11. Annual Report 1800, 8880/17, ARM: "Los títulos de sus pertenencias estaban por la mayor parte obscurecidos entre el polvo, y sin saberse si se habían perdido sus emolumentos y desde qué tiempo se había dejado de hacer."

3. Annual Report 1799, 8880/17, ARM: "Poniendo bajo nuestra dirección todo lo gubernativo de la Casa, cual es la administración de la mencionada Inclusa, cuidase de los Niños de ella, del ingreso de sus caudales y Rentas, quedando el Juez Protector únicamente para conocer de los Pleitos o derechos que puedan pertenecerla, pues ha de correr enteramente y en todo su Rango a cargo de la Junta."

4. See, for instance, the statutes of the female society for taking care of the foundlings in Málaga, which forbade the women from intervening in the finances of the house under any circumstances: "En ningún tiempo se podrá la misma Asociación, ni Señora alguna, introducir, ni intervenir en el manejo de caudales de la Casa de Expósitos, ni en todo ni en parte de ello, aun cuándo se lo supliquen." In *Real Orden de S.M por la que se sirve aprobar los Estatutos de la Asociación de Señoras establecida en esta Ciudad para ejercitar la caridad en el cuidado de los Niños Expósitos* (Málaga: Imprenta de Luis de Carreras y Ramón, 1796).

5. Alan Forrest, "Poverty," in *The Oxford Handbook of the Ancient Régime*, ed. William Doyle (Oxford: Oxford University Press, 2012), 167–82, here 179. See also Sandra Cavallo, "The Motivations of Benefactors: An Overview of Approaches to the Study of Charity," in *Medicine and Charity before the Welfare State*, ed. Jonathan Barry and Colin Jones (London: Routledge, 1994), 46–62; on the relations between charity, the state, and the economy, see Jonathan Barry

and Colin Jones, "Introduction" in the same volume, pp. 1–13. The relationship between population policies, sexual freedom, and foundling houses is discussed in Faramerz Dabhoiwala, *The Origins of Sex: A History of the First Sexual Revolution* (London: Penguin Books, 2013), 215–50. See also Julie Hardwick, *Sex in an Old Regime City: Young Workers and Intimacy in France, 1660–1789* (Oxford: Oxford University Press, 2020).

6. Contemporary denunciations of the situation of Spanish foundling houses can be found in Antonio Bilbao, *Destrucción y conservación de los expósitos: Modo breve de poblar la España* (Málaga: Don Félix de Casas y Martínez, 1790); Pedro Joaquín de Murcia, *Discurso político sobre la importancia y necesidad de hospicios, casas de expósitos y hospitales que tienen todos los estados y particularmente España* (Madrid: Imprenta Viuda de Ibarra, 1798); and Juan Antonio Trespalacios y Mier, *Discurso sobre que los niños expósitos consigan en las inclusas el fin de estos establecimientos* (Madrid: Imprenta de Villalpando, 1798). See also the regulations approved by the Spanish Crown in 1796: *Real Cédula de S. M. por la que se manda observar el Reglamento inserto para la Policía general de Expósitos de todos sus dominios*. Available online as "Cédula 1796-12-11" at http://www.liburuklik.euskadi.eus/handle/10771/26927.

7. Valerie Fildes, *Wet Nursing: A History from Antiquity to the Present* (New York: Basil Blackwell, 1988), 145–57. Mortality rates varied greatly annually. See idem., 170.

8. Rachel Ginnis Fuchs, *Abandoned Children: Foundlings and Child Welfare in Nineteenth-Century France* (Albany: State University of New York Press, 1984), 19–27. Olwen H. Hufton provides numbers of abandoned children annually in French cities: about eight thousand were abandoned in Paris ca. the 1780s; two thousand in Lyon; six hundred in Bordeaux; and three hundred in Dijon: Olwen H. Hufton, *The Poor of Eighteenth-Century France 1750–1789* (Oxford: Oxford University Press, 1974), 332. On the London Foundling Hospital, see: Alysa Levene, *Childcare, Health and Mortality at the London Foundling Hospital, 1741–1800: "Left to the Mercy of the World"* (Manchester: Manchester University Press, 2007).

9. "Libro de Juntas que mandó celebrar el actual protector de la Casa de la Inclusa Gonzalo Josef as," 5109/2, 60–64, ARM. This notebook contains the minutes of meetings of the board of the Madrid Foundling House before the Junta. See especially January 16 (order by Godoy for allowing a commission of the Junta to visit the foundling house); January 24; February 12; February 23; and March 4 (the Junta asks for information on the last five years about the intake of children). See also Paula Demerson, "La Real Inclusa de Madrid a finales del siglo XVIII," *Anales del Instituto de Estudios Madrileños* 8 (1972): 261–72.

10. See chapter 1. The list of all the members is found in Theresa Anne Smith, *The Emerging Female Citizen: Gender and Enlightenment in Spain* (Berkeley: University of California Press, 2006), 137–39, table 5.

11. Antonio Carreras Panchón, "La mortalidad en las inclusas españolas (1800–1808)," *Cuadernos de Historia de la Medicina Española* 14 (1975): 261–68; idem., *El problema del niño expósito en la España Ilustrada* (Salamanca: Universidad de Salamanca, 1977). See also Vicente Pérez Moreda, *La infancia abandonada en España (siglos XVI–XX)* (Madrid: Real Academia de la Historia, 2005). In Spain, the doctor Ignacio María Ruiz de Luzuriaga compiled the numbers from Spanish foundling houses: see Ignacio María Ruiz de Luzuriaga, *Estadística politico mé*dica (Mss., 1817–) vol. 1, 39–62, *Signatura: 21–3ª Biblioteca/15*, Archivo Real Academia de Medicina de Madrid (RAM). See also *Expediente sobre los niños expósitos y los hospicios* (1793), Mss/11267/32, Biblioteca Nacional de España (BN).

12. In 1790 the expenses amounted 95,187 reals; in 1800, after the Junta took over, this increased to 156,059 reals; and in 1805, expenses peaked at 264,415 reals (Sherwood, *Poverty*, 47).

A detailed analysis of the incomes and expenses of the Madrid Foundling House can be found in Sherwood, *Poverty*, 4–33.

13. The old Madrid Foundling House was on calle Preciado. On September 2, 1801, it moved to El Soldado 3–4 (abutting today's calle de la Libertad). On the new venue, see the poem by Francisco Gregorio Salas, "La nueva Inclusa," Mss B. 106, Colección de papeles curiosos en verso y prosa, tomo IV, 147, Biblioteca de las Cortes Españolas. Quoted in Demerson, *María Francisca de Sales*, 219.

14. Theodore M. Porter, "Making Things Quantitative," *Science in Context* 7, no. 3 (Autumn 1994): 389. See also Ted Porter, *Trust in Numbers: The Pursuit of Objectivity in Science and Public Life* (Princeton, NJ: Princeton University Press, 1995); Ulrich Tröhler, "Quantifying Experience and Beating Biases: A New Culture in Eighteenth-Century British Clinical Medicine," in *Body Counts: Medical Quantification in Historical and Sociological Perspective*, ed. Gérard Jorland, Annick Opinel, and George Weisz (London: McGill-Queen's University Press, 2005), 19–50.

15. Andrea Rusnock, "Quantifying Infant Mortality in England and France, 1750–1800," in Jorland, Opinel, and Weisz, *Body Counts*, 65–86, here 66.

16. Valentina Fernández Vargas, "Informe sobre el archivo de la Antigua Inclusa de Madrid, hoy Instituto Provincial de Puericultura," in Pedro Espina Pérez, *Historia de la Inclusa de Madrid* (Madrid: Defensor Menor Comunidad de Madrid, 2005), 47–57. The oldest *Libro de Entradas* kept runs from 1588 to 1589. The index is by alphabetical order of first names.

17. See the discussion of this term in the introduction to this book.

18. Frances Miley and Andrew Read, "Go Gentle Babe: Accounting and the London Foundling Hospital, 1757–1797," *Accounting History* 21, nos. 2–3 (May 2016): 167–84. Also Ruth K. McClure, *Coram's Children: The London Foundling Hospital in the Eighteenth-Century*. New Haven: Yale University Press, 1981. On the relation to the Hôpital des Enfants Trouvés, see Fuchs, *Abandoned Children*, 19–27.

19. Ben Kaftka, "Paperwork: The State of the Discipline," in *Book History*, 12 (2009): 340–53; idem., "Hunting the Plumed Mammal: The History of 'Bureaucracy' in France, 1750–1850," in *Figures of Authority: Contributions Towards a Cultural History of Governance*, ed. Peter Becker and Rüdiger von Krosigk (Brussels: Peter Lang, 2008), 119–26; Cornelia Vismann, *Files: Law and Media Technology*, abridged and translated by Geoffrey Winthrop-Young (Stanford, CA: Stanford University Press, 2008); Jon Agar, *The Government Machine: A Revolutionary History of the Computer* (Cambridge, MA: MIT Press, 2003); Valentin Groebner, *Who Are You? Identification, Deception, and Surveillance in Early Modern Europe*, trans. Mark Kyburz and John Peck (New York: Zone Books, 2007); Peter Becker and William Clark, "Introduction," in *Little Tools of Knowledge: Historical Essays on Academic and Bureaucratic Practices*, ed. Peter Becker and William Clark (Ann Arbor: University of Michigan Press, 2001), 1–35.

20. Francisco Sánchez-Blanco, *El absolutismo y las luces* (Madrid: Marcial Pons, 2002), 340–48 and 371–423. See also several articles in Enrique Fuentes Quintana, ed., *Economía y economistas españoles: La Ilustración* (Barcelona: Galaxia Gutenberg, 2000), including Enrique Fuentes Quintana, "Una aproximación al pensamiento económico de Jovellanos a través de las funciones del Estado" (331–420); Vicent A. Llombart Rosa and Jesús Astigarraga Goenaga, "Las primeras 'antorchas de la economía': Las sociedades económicas del país en el siglo XVIII"; and Ernest Lluch and Salvador Almenar Palau, "Difusión e influencia de los economistas clásicos en España (1776–1870)."

21. Letter from the Countess of Montijo to the Junta, *Proposición de la condesa de Montijo*

sobre lo útil que sería el que solicitase tomar a su cuidado la crianza de los niños expósitos de la Inclusa de Madrid in Archivo de la Sociedad Económica Matritentense (Archivo de la Real Sociedad Económica Matritentse, SEM), leg. 92–92 (July 3, 1789): "Los hombres gobiernan con sus talentos y sus luces las repúblicas, la manera en que las mujeres pueden contribuir a la pública felicidad es cuidando a esas criaturas inocentes." See also Demerson, *María Francisca de Sales*, 201–13.

22. On this point, see the discussion in the introduction.

23. Annual Report for the year 1796, 8487/2, ARM: "las inocentes víctimas del desorden o de la excesiva pobreza de sus padres, nos tienden los brazos y reclaman nuestra protección, nuestro cuidado."

24. Sylvana Tomaselli, "Moral Philosophy and Population Questions in Eighteenth-Century Europe," *Population and Development Review* 14, supplement (1988): 7–29.

25. Letter of the Countess of Montijo to the Junta, leg. 105/6, SEM: "excede bastante más de la mitad el número de expósitos que se pierden antes de cumplir el año."

26. The report gives no information on the source of these "tables of general mortality."

27. Letter of the Countess of Montijo to the Junta, leg. 105/6, SEM: "pérdida tan exorbitante como lastimosa que pide la más seria atención de toda Sociedad Patriótica, para averiguar y remedial las causas que la producen."

28. Letter from the Countess of Montijo to Gonzalo Josef Vilches, March 2, 5109/2, ARM: "Carta de la Condesa de Montijo al protector para que le facilite la entrada/ salida de niños en los 5 últimos años: los niños que entran en ella anualmente, cuantos se crían en la misma casa, que número de amas mantiene en ella, cuantos fuera en Madrid, que número se cría en los lugares, que número se prohíjan, cuantos fallecen y cuantos llegan a entrar en la casa de los Desamparados."

29. Annual Report of 1798, 8880/17, ARM.

30. Children were sent to the Casa de los Desamparados when boys turned eight and girls seven.

See *Razón de las criaturas en la sala de Amas correspondientes al mes de Octubre de 1799*, 8456/2, ARM.

31. Report on the number of children received by the foundling house, *Razón de las criaturas que han entrado en la Real Inclusa de esta corte en este presente mes de marzo de 1801*, 8456/2, ARM.

32. Annual Report of 1800, 8880/17, ARM.

33. The sisters also worked in the Paris Enfants Trouvés Hospital. Quoted in Joan Sherwood, *Infection of the Innocents: Wet Nurses, Infants and Syphilis in France* (Montreal: McGill-Queen's University Press, 2010), 6. See also *Verdadera y sucinta noticia de las ocupaciones y buenos servicios que hacen a la humanidad las Hijas de la Caridad* (Madrid: s.n., 1813). An exemplar can be found in 881/13, ARM. The Junta first contracted six nuns, and then another six in 1801. Quoted in Demerson, *María Francisca de Sales*, 220. See also *Libro de Actas de 1800 y 1801*, A/56–57. SEM: Junta, April 17: "It was agreed to bring four nuns from the village of Barbastro to the Madrid Foundling House, and to paint the cribs and the beds of the wet nurses."

34. Draft of the contract with the nuns, *Borrador de los artículos para formar la contrata a las hermanas*, 8881/14, ARM.

35. The notebooks of kitchen expenses are in folder 8881/16, ARM, with the expenses of nuns, wet nurses, and weaned children.

36. Sherwood, *Poverty*, 5, 1–33, and 33–48.

37. Annual Report from year 1801, 8880/17, ARM; *Papel presentado por la Montijo a la Junta*

de Damas de 1800, 8342, ARM; and *Instrucciones de la Condesa de Montijo para habilitar una sala arriba para niños enfermos*, 8355, ARM.

38. Sherwood, *Infection of the Innocent*, 17; Fildes, *Wet Nursing*, 71–72; idem., *Breast, Bottles and Babies: A History of Infant Feeding* (Edinburg: Edinburg University Press, 1986); 329n1. For a discussion of contemporary thinking on breastfeeding and contagion, see http://www.scielo.br/scielo.php?pid=s0021-75572004000700010&script=sci_arttext&tlng=en.

39. Fildes, *Wet Nursing*, 111–25. On the Spanish context, see Mónica Bolufer Peruga, "Actitudes y discursos sobre la maternidad en la España del siglo XVIII: La cuestión de la lactancia," *Historia Social* 4 (1992): 3–22.

40. George D. Sussman, *Selling Mother's Milk: The Wet-Nursing Business in France 1715–1914* (Urbana: University of Illinois Press, 1982); Jutta Gisela Sperling, *Medieval and Renaissance Lactations: Images, Rhetoric, Practices* (Burlington, VT: Ashgate, 2013), 10–11. Sussman (*Selling Mother's Milk*, 73–79) analyzed the letters exchanged by a family of watchmakers and their wet nurse in the country. See also Fildes, *Wet Nursing*, 159–80.

41. For the wet-nursing market in Madrid, see Carmen Sarasúa, *Criados, nodrizas y amos: El servicio doméstico en la formación del mercado de trabajo madrileño 1758–1868* (Madrid: Siglo Veintiuno de España Editores, 1994), 158. See Antonio Espinosa, *El ceremonial de estrados y critica de visitas: Obra util, curiosa, y divertida* (Madrid: Antonio Espinosa, 1789), chapter 3.

42. *Libro de Entradas de Criaturas* 168 (1800), 9053/1, 73, ARM.

43. In fact, according to the former administrators, there were many wet nurses who intended to commit the "trick" of abandoning their child and later seek a job in the foundling house with the hope of receiving their own child. Quoted in Luzuriaga, *Estadística politico m*édica. On Silberto, see *Libro de Salidas de Criaturas* 168 (1800), 9215, 180, ARM: Silberto was sent to San Sebastián de los Reyes four days later with the wet nurse Lucía Tomasa, on January 28, 1800. He was returned on August 28 and died in the infirmary of the house on September 2, 1800.

44. Sherwood, *Infection of the Innocent*, 9–10; see Fildes, *Wet Nursing*, 101–10 on the occupational diseases of wet nurses.

45. Sherwood, *Infection of the Innocents*, 9; Fildes, *Wet Nursing*, 71–73.

46. Jaume Bonells, *Perjuicios que acarrean al género Humano y al Estado las Madres que rehusan criar a sus hijos, y medios para contener el abuso de ponerlos en Ama* (Madrid: Miguel Escribano, 1786), 351.

47. Quoted in Luzuriaga, *Estadística político m*édica: "Las nodrizas que lactan en esta casa son de reprobada conducta . . . se juzgan tan precisas y necesarias en los tiempos en que abundan los expósitos que no hay libertad para corregirlas; se conchaban entre sí . . . en cuyo conflicto tiene el Administrador que armarse de paciencia, concediéndoles cuanto quieren."

48. Elena Serrano, "Sex and Prisons: Women and Spanish Penitentiary Reform, 1787–1808," in "The Sexes and the Sciences," edited by D. N. Wagner and J. Wharton, special issue of *Journal for Eighteenth-Century Studies* 42, no. 2 (December 2019): 501–17.

49. "Reglamento de las Hermanas de la Caridad," 1799, ARM, 164/11.

50. Letter of the Junta to Tomas Mas, July 30, 1800, 8476/2, ARM.

51. The women of the Junta were well aware of Genlis's works on education. They subscribed to the Spanish translation of Genlis, *Les Veillées du château* (Paris: M. Lambert et F.-J. Baudouin), which appeared in Spanish in 1788: de Genlis, *Veladas de la quinta o Novelas e historias sumamente útiles para que las madres de familia, a quienes las dedica la autora, puedan instruir á sus hijos, juntando la doctrina con el recreo* (Madrid: M. González, 1788).

52. Mme de Genlis, "Letter XXI from the Baronesse d'Almane to Mm. d'Ostalis," in *Adelaide and Theodore or Letters in Education*, ed. T. G. Dow (1783). Quoted in Margaret Carlyle, "Breastpump Technology and 'Natural' Motherly Milk in Enlightenment France," *Women's Studies International Forum* 60 (January–February 2017): 89–96.

53. Sherwood, *Poverty*, 12–17.

54. Luzuriaga, *Estadística politico médica*.

55. Draft of the contract to the nuns, *Borrador a los artículos para otorgar la contrata de las hermanas de la caridad*, 8881/14, ARM: "La hermana de sala tendrá el mayor cuidado de que las amas vistan los niños en las horas señaladas, con el orden, cariño y silencio convenientes, advirtiéndoles lo que han hecho con poco cuidado o con precipitación, como si alguno está oprimido, o mal envuelto o poco limpio."

56. See the correspondence of the Junta with priests in 8416/014, ARM. The Junta offered to send thirty-eight hundred certifications of "good behavior" to the Bishop of Santa María, for priests to give to the wet nurses who deserved them.

57. Notebook, *Registro de Amas de Cría desde 1803*, 10300/2, ARM. This is a notebook where the names of the wet nurses are listed along the cribs for which they are responsible. For instance, "Ama Faustina Muñoz, children in 15 and 16." This is followed by the list of children who occupied the cribs and their number in the *Libro de Entrada y Salida de Criaturas*. It seems that in the lactation wards (Sala de Amas) there were sixty-six cribs in 1803—this is the largest crib number in the notebook. The wet nurses in the infirmary and their salaries as well as other workers' salaries are found in *Libro de Tandas*, desde 1 Enero 1800 hasta 31 Dic. de 1826, ARM, 10317/4.

58. Report, *Plan de las tareas de la Junta de Señoras*, leg. 145–8, SEM. Quoted in Demerson, *María Francisca de Sales*, 217 and footnote 4: "sus virtudes sociales y caseras, aquellas virtudes que esparcen la alegría y aprietan los lazos de las familias, han sido unas lecciones vivas que han recibido y reciben diariamente las amas y demás dependientes de la casa."

59. Dolores Palma García, "Las escuelas patrióticas creadas por la Sociedad Económica Matritense de Amigos del País en el siglo XVIII," *Cuadernos de Historia Moderna y Contemporánea* 5 (1984): 37–56.

60. "Libro de Actas de la Junta de damas, 1795–1796," May 8, 1795: Libro A/56–4, SEM. See also the description on the volumes that were to be presented to the queen, written by the Marquise of Ariza. These volumes were given by the Countess de O'Reilly and the Marquise of Sonora.

61. Condesa de Castro-Terreño, *Elogio de la reyna nuestra señora* (Madrid: Imprenta Real, [1801?]), 30–31: "Que se miraba como corriente el que estas criaturas racionales quitasen a los perros el ministerio fatal de extraer con sus labios inocentes leches envenenadas, y descargar pechos llenos de un humor mortal." See also Bonells, *Perjuicios*, 350–53. The exemplar in the Biblioteca Nacional de Madrid belonged to the library of the Duchess of Osuna. The excess of milk was thought to be dangerous and in need of removal by mechanical methods, such as suckling glasses and breast pumps. On breast pumps, see Carlyle, "Breast-Pump Technology and 'Natural' Motherly Milk," 89–96.

62. Castro-Terreño, *Elogio*, 34: "Noventa y seis niños por ciento morían antes de que la piedad de la Reina los pusiese al cuidado maternal de la Junta: esta mortandad horrible fue progresivamente cediendo a las nuevas mejoras: en el mes ultimo bajó a cuarenta y dos."

63. Josefa Díez de la Cortina y Morales, *Elogio de la Reyna Nuestra Señora . . . leído en la junta publica de distribución de premios en 18 de diciembre de 1799* (Madrid: Imprenta Real, 1800):

"La economía sucede al desorden, la inteligencia a la ignorancia, un cuidado activo y celoso a la negligencia mercenaria. . . . La humanidad se felicita de haber concebido un proyecto tan interesante: el estado contará nuevos individuos y las artes y oficios adquirirán nuevos brazos."

64. Josefa Gómez de Enterría, *Voces de la economía y el comercio en el español del siglo XVIII* (Alcalá: Universidad de Alcalá, 1996); Juana Ugarte Blanco, "Economía en Espagne au 18e siècle: Fonctionnements discursifs et sociaux," *Mots* 15 (1987): 93–109.

65. Lissa Roberts, "Practicing Oeconomy during the Second Half of the Long Eighteenth Century: An Introduction," *History and Technology* 30, no. 3 (2014): 133–48. For a discussion of the practices in Spain, see Elena Serrano, "Making Oeconomic People: The Spanish Magazine of Agriculture and Arts for Parish Rectors (1797–1808)," *History and Technology* 30 (2014): 149–76. See also Lissa Roberts and Simon Werrett, "Introduction: A More Intimate Acquaintance," in *Compound Histories: Materials, Governance and Production, 1760–1840*, ed. Roberts and Werrett (Leiden: Brill, 2017), 1–32.

66. Simon Werrett, *Thrifty Science: Making the Most of Materials in the History of Experiment* (Chicago: Chicago University Press, 2019). See also idem., "Household Oeconomy and Chemical Inquiry," in Roberts and Werrett, *Compound Histories*, 35–56; idem., "The Sociomateriality of Waste and Scrap Paper in Eighteenth-Century England," in *Working with Paper: Gendered Practices in the History of Knowledge*, eds. Carla Bittel, Elaine Leong, and Christine Von Oertzen (Pittsburgh: University of Pittsburgh Press, 2019), 46–59.

67. I quote the Spanish version of François de Salignac de La Mothe Fénelon's *Traité de l'éducation des filles*, as *Tratado de la educación de las hijas* (Madrid: Viuda de Eliseo Sánchez, 1769), published in English as *Instructions for the Education of a Daughter* (London: Printed by and for J. Ferraby).

68. Josefa Amar y Borbón, *Discurso sobre la educación física y moral de las mujeres* (Madrid: Benito Cano, 1790), 158.

69. On the statutes, see Olegario Negrín Fajardo, *Ilustración y educación: La Sociedad Económica Matritense* (Madrid: Editora Nacional, 1984), 7–73.

70. On the contemporaneous importance of baptism, see the translation carried out by the secretary of the Junta, the Countess of Montijo: Nicolas Letourneux, *Instrucciones cristianas sobre el sacramento del matrimonio* (Barcelona: Bernardo Pla, 1774), 116–215. The officials of the house carefully checked the proofs of baptism, and in dubious cases, they preferred to baptize again. See, for instance, the discussion on the veracity of the baptism certification brought by the child Serapio Miguel Menoy: 7721/7, AHR.

71. Luzuriaga, *Estadística*.

72. Crisanto de Bazo brought a letter from a clergyman with him testifying that the family's circumstances had forced them to abandon their child and that they wanted to have her back. 9054/1, ARM, 326.

73. Parchments inside the pages of the *Libros de Entradas de Criaturas*. See, for instance, the parchments in *Libro de Entradas* for the year 1804 (172): 7721/9, ARM. See also, in 7721/1, ARM, parchments along with certificates from the parish rector certifying the good behavior (*certificados de buena conducta*) of wet nurses in their parishes.

74. *Libro de Salidas de Criaturas*, 169, 9216/1, ARM, 149.

75. Luis Luque, *Arte de partida doble ilustrado* (Cádiz: Calle de la Compañía, 1783).

76. Luque, *Arte de partida doble*, 34.

77. Note also the connections between mercantile and scholarly bookkeeping in the practices of world travelers (especially in botany). Anke te Heesen, "Accounting for the Natural

World: Double-Entry Bookkeeping in the Field," in *Colonial Botany: Science, Commerce, and Politics in the Early Modern World*, ed. Londa Schiebinger and Claudia Swan (Philadelphia: University of Pennsylvania Press), 237–51.

78. Anke te Heesen, "The Notebook: A Paper-Technology," in *Making Things Public: Atmospheres of Democracy*, ed. Bruno Latour and Peter Weibel (Cambridge, MA: MIT Press, 2005), 582–89.

79. See, for example, *Constituciones, ordenanzas y reglamentos del Real Hospicio . . . de la Ciudad de Badajoz* (Madrid: Josef Collado, 1804), 61–64.

80. Thanks to Maria Rentetzi for her insightful comment about this.

81. See the letter that the Junta sent to D. Josef de Santa María, 8416/14, ARM.

82. Annual Roport of 1801, 8880/17, ARM.

83. Annual Report of 1802, 8888/17, ARM.

84. "Comunicaciones emitidas y/o recibidas por la Junta de Damas de Honor y Mérito sobre disposiciones dadas a los oficiales de libros del Despacho de la Inclusa y Colegio de la Paz, 1802-1831," 8342/009, ARM.

85. Books of deceased children, 10264/2, ARM; and 1064/3, ARM. The first one is from 1802 to 1808 and the second one from 1809 to 1812.

86. Annual Report 1806, 8880/17, ARM.

87. Porter, "Making Things Quantitative," 399; idem., "Funny Numbers," *Culture Unbound: Journal of Current Cultural Research* 4 (2012): 585–98.

88. Francesca Bray, *Technology, Gender and History in Imperial China: Great Transformations Reconsidered* (London: Routledge, 2013), 17.

Chapter 3. Testing Baby Food

1. Sociedad Económica Matritense (SEM), 164/10: "Por la abundancia de criaturas y escasez de las Amas, les fue preciso valerse del uso de las cabras puestas en un Potro con un colchón debajo donde cómodamente maman dos niños de un solo animal, sin haberse experimentado el menor contratiempo."

2. Santiago García, *Instituciones sobre la crianza física de los niños expósitos* (Madrid: Imprenta de Vega y Compañía, 1805).

3. Ignacio María Ruiz de Luzuriaga, *Estadística político médica: Estado comparativo de los xenodochios, derephotrofios y horfanotrofios*, 1817–1819 (Mss: Real Academia de Medicina de Madrid, RAM, 21–3a Biblioteca/15–19): In cases where wet nurses had to care for three children, they were to use paps (soft food, mainly made with bread and water or broth or milk) and goat's milk.

4. García, *Instituciones*, 74–81. García listed "asta de vaca, vejiga, cuchara, biberón o pistero, and muñequitas."

5. Valerie Fildes, *Wet Nursing: A History from Antiquity to the Present* (New York: Basil Blackwell, 1988), 153–58; idem., *Breast, Bottles and Babies: A History of Infant Feeding* (Edinburg: Edinburgh University Press, 1986); Ian G. Wickes, "A History of Infant Feeding, Part III: Eighteenth and Nineteenth-Century Writers," in *Archives of Disease in Childhood* 139 (June 1953): 332–40.

6. Letter from Pedro Lasplanas, April 27, 1812: 8342/7, Archivo Regional de Madrid (ARM): "En la tarde de este día he diseccionado un niño, de los que se habían alimentado con la leche de cabras."

7. Paula Demerson, *María Francisca de Sales y Portocarreño, condesa del Montijo: Una figura*

de la Ilustración (Madrid: Editora Nacional, 1975). The Countess of Montijo was secretary of the Junta from its initial founding in 1787 until her exile in 1805, with a brief interruption of several months between 1804 and 1805. See also Paloma Fernández Quintanilla, *La IX Duquesa de Osuna: Una Ilustrada en la Corte de Carlos III* (Madrid: Ediciones 12 Calles, 2017), 350–53.

8. Joan Sherwood, *Poverty in Eighteenth-Century Spain: The Women and Children of the Inclusa* (Toronto: Toronto University Press, 1988).

9. Sherwood, *Poverty*, 151.

10. Sherwood, *Poverty*, 195.

11. Sherwood, *Poverty*, 155.

12. Elaine Leong and Alisha Rankin, "Testing Drugs and Trying Cures: Experiment and Medicine in Medieval and Early Modern Europe," in "Testing Drugs and Trying Cures," special issue, *Bulletin of the History of Medicine* 91, no. 2 (Summer 2017): 157–82.

13. Elaine Leong, *Recipes and Everyday Knowledge: Medicine, Science and the Household in Early Modern England* (Chicago: University of Chicago Press, 2018), 4.

14. See, for instance, Justin Rivest on tests done for one febrifuge drug in three French naval hospitals that employed five hundred doses; Justin Rivest, "Testing Drugs and Attesting Cures: Pharmaceutical Monopolies and Military Contracts in Eighteenth-Century France," in "Testing Drugs and Trying Cures," special issue, *Bulletin of the History of Medicine* 91, no. 2 (Summer 2017): 362–90.

15. Emma Spary, *Feeding France: New Sciences of Food, 1760–1815* (Cambridge, UK: Cambridge University Press, 2014), 8–15.

16. Spary, *Feeding France,* especially chs. 2, 5, and 6; Anna Maerker, "Political Order and the Ambivalence of Expertise: Count Rumford and Welfare Reform in Late Eighteenth-Century Munich," in "Expertise and the Early Modern State," ed. Eric H. Ash, special issue, *Osiris* 25, no. 1 (2010): 213–30; Rebecca Earle, "Potatoes and the Hispanic Enlightenment," *Americas* 75, no. 4 (Winter 2018): 639–60; see also chapter 4 of this book.

17. Rebecca Earle, "Promoting Potatoes in Eighteenth-Century Europe," *Eighteenth-Century Studies* 51, no. 2 (Winter 2017): 147–62, here 148.

18. See chapter 2 of this book, especially the section "Mothering Foundlings."

19. Pascual Mora, *El hombre en la primera época de su vida*, 3 vols. (Madrid: Martínez Dávila, 1827), 1:237–38. Mora was in turn quoting another famous Spanish doctor, José Iberti.

20. Nora Doyle, *Maternal Bodies: Redefining Motherhood in Early America* (Chapel Hill: University of North Carolina Press, 2018), 116–45. On nipple shields, see Wickes, "History of Infant Feeding," 337, which mentions those designed by Mrs. Relf, sold in the shop of the surgical instrument maker Mr. Savigny in Covent Garden. On the breast pump, see Margaret Carlyle, "Breast-Pump Technology and 'Natural' Motherly Milk in Enlightenment France," *Women's Studies International Forum* 60 (January–February 2017): 89–96. See also William Rowley, *A Practical Treatise on Diseases of the Breasts of Women . . . A Plate for a New Invented Machine, for More Commodiously Drawing the Milk from the Breasts* (London: Newbery, 1772). On motherhood and lactation in Spain, see Mónica Bolufer Peruga, "Actitudes y discursos sobre la maternidad en la España del siglo XVII: La cuestión de la lactancia," *Historia Social* 14 (1992): 3–22.

21. Fildes, *Breast, Bottles and Babies*, 262–306. For the situation in Spain, see Pedro Saura, "Tratado y explicación de la teta artificial y modo de criar los niños sin amas," *Semanario Económico* 13 y 20-VIII (1767), quoted in Carmen Labrador and J. Carlos de Pablos, *La Educación en los papeles periódicos de la Ilustración española* (Madrid: Ministerio de Educación y

Ciencia, 1989). See also George Armstrong, *An Account of the Diseases Most Incident to Children* (London, T. Cadell, 1771). For a discussion of the English context, see T. G. H. Drake, "Infant Feeding in England and in France from 1750 to 1800," *American Journal of Diseases of Children* (1930): 1049–61.

22. George D. Sussman, *Selling Mother's Milk: The Wet-Nursing Business in France 1715–1914* (Urbana: University of Illinois Press, 1982), 79–93. Sussman's book quotes Rolands's *Avis a ma fille*.

23. The Marquise of Llano, Isabel Parreño Arce Ruiz de Alarcon y Valdes. Her husband José Agustín Llano y de la Cuadra (1722–1794) was appointed ambassador to Vienna on October 26, 1785, where he moved the next year. The Marquise of Llano was admitted as a member of the Junta on February 8, 1788; see Theresa Ann Smith, *The Emerging Female Citizen: Gender and Enlightenment in Spain* (Berkeley: University of California Press, 2006), 136, table 5.

24. Letter from the Marquis del Llano to the Countess of Montijo, December 9, 1788, SEM. Junta de damas 100/11: "El cual ha adoptado muchas de las Damas para criar a sus hijos prefiriéndolo a las Amas de leche."

25. Josefa Amar y Borbón, *Discurso sobre la educación física y moral de las mujeres* (Madrid: Imprenta de Benito Cano, 1790).

26. Josefa Amar was made honorary member of the Barcelona Royal Academy of Medicine in 1790. "Libro de Acuerdos y resoluciones de la Academia Médico Práctica de Barcelona," vol. 1 (1770–1795), 218–19, Arxivo de la Reial Acadèmia de Medicina de Catalunya. These notebooks of the academy's minutes were kept in several volumes.

27. "Libro de Actas de la Junta de damas, 1795–1796," Libro A/56-4, February 19, 1795, SEM: "La presidenta leyó un papel proponiendo a la Junta convendría se establecieran las dos comisiones de educación que previene el título 8°de los Estatutos." These notebooks contained the minutes of the Junta's meetings (Hereafter, "Libro de Actas de la Junta"). A discussion of the contents of these reports is found in Smith, *Emerging Female Citizen*, 150–57. A list of the queries is found in table 6 of the same book.

28. "Informe sobre sobre el mérito de las siete memorias formadas por las señoras," March 14, 1801, 146-11, SEM. This report summarized the accounts penned by seven members of the Junta.

29. Quoted in Christine Adams, *A Taste for Comfort and Status: A Bourgeois Family in Eighteenth-Century France* (University Park: Pennsylvania State University Press, 2000), 205. The manuscripts of the trials are held in in the Bibliothèque Municipale, Bordeaux.

30. The experiments ran from July 20 to August 17, 1783. See Adams, *Taste for Comfort*, 206; see also 172, 180, and 205–7. On the early years of the London Foundling Hospital, see Fildes, *Wet Nursing*, 153–18 and 161.

31. Quoted in Fildes, *Wet Nursing*, 146–47 and 161. See Colin Jones, *Charity and Bienfaisance: The Treatment of the Poor in the Montpellier Region 1740–1815* (Cambridge, UK: Cambridge University Press, 1982), and for England, *An Account of the Hospital for the Maintenance and Education of Exposed and Deserted Young Children* (London: Publishing house unknown, 1749), 8–10; V. E. Lloyd Hart, *John Wilkes and the Foundling Hospital at Aylesbury 1759–1768* (Aylesbury: HM & M, 1979), 16–23.

32. Its statutes are found in *Real Orden de S.M. por la que se sirve aprobar los estatutos de la Asociación de Señoras establecida en esta ciudad para ejercitar la caridad en el cuidado de Niños Expósitos* (Málaga: Imprenta de Luis de Carreras y Ramón, 1796): "Serán muy precavidas en formar Proyecto de crianza física de los niños, de si se puede con cabras, si con leche de vacas,

si con pechos artificiales, por ser mucho lo que en este punto se ha hablado y escrito en todos los tiempos."

33. Juan Antonio Trespalacios y Mier, *Discurso sobre que los niños expósitos consigan en las inclusas el fin de estos establecimientos* (Madrid: Imprenta de Villalpando, 1798).

34. Trespalacios y Mier, *Discurso*, 14: "Porque a la verdad ellas solas son las que mejor lo entienden, y las que adquieren noticias mas verídicas de la sanidad y costumbres de las mismas nodrizas, de la calidad de las leches, y las que mejor saben también discernir si es útil en algunos países el uso de las leches de cabras, ú otros irracionales, o es efectivamente perjudicial como parece: y finalmente evitar asimismo el abuso intolerable de las adormideras, del pan mojado, y otros no menos perjudiciales a estos infelices."

35. Amar y Borbón, *Discurso sobre la educación*, 36–40. An insightful discussion on the role of the "maternal body's matter" and the theory of humors can be found in Victoria Sparey, "Identity-Formation and the Breastfeeding Mother in Renaissance Generative Discourses and Shakespeare's Coriolanus," *Social History of Medicine* 25, no. 4 (October 2012): 777–94.

36. Amar y Borbon, *Discurso sobre la educación*, 38–39.

37. Drake, "Infant Feeding," 152.

38. Quoted in Wickes, "History of Infant Feeding," 337. See also Fildes, *Wet Nursing*, 146.

39. García, *Instituciones*, 72–73.

40. Barbara Orland, "Enlightened Milk: Reshaping a Bodily Substance into a Chemical Object," in *Materials and Expertise in Early Modern Europe: Between Market and Laboratory*, ed. Ursula Klein and Emma C. Spary (Chicago: University of Chicago Press, 2010), 163–98. See also Wickes, "History of Infant Feeding," 337, where he mentions the analysis of women's milk by Michel Underwood in his *Treatise on Diseases of Children* (1799) and the discussions with his contemporary Dr. Clarke and others.

41. Samuel Ferris, *A Dissertation on Milk* (London: John Abraham, 1785), 14–24.

42. Suggestion of the doctor Santiago García for increasing the number of wet-nurses in the foundling house, January 3,1800, 8434/5, ARM: "El ensayo que actualmente se está hacienda con los que se hallan en poder de dos mujeres que a sus expensas ha puesto las Excelentísimas Señoras de Truillas y Sonora, dan una esperanza bien fundada."

43. "Memoria que contiene los trabajos y ocupaciones de la Junta de Señoras de Honor y Mérito de 1803," 176–79, SEM: "La casa del Duque de Granada contigua al convento de San Bernardino, para que se hicieran la lactancia de los niños con leche de cabras, según el medico Santiago García." These "memorias" were annual reports penned by the director of the Junta, to be sent to the Madrid Economic Society; the title could be translated as "Report of the Works and Business of the Junta de Señoras de Honor y Mérito of 1803." I will hereafter refer to these documents as "Memoria Anual de la Junta."

44. "Libro de Actas de la Junta," February 18, 1805, SEM. See also letter from the Countess of Montijo, June 28, 1805, 999, RAM; letter from Ignacio M. Ruíz de Luzuriaga to the Countess of Montijo about a book by Santiago García, August 22, 1805, 999, RAM; letter from Tomás García Suelto, May 1803, 1270, RAM.

45. Letter from the Countess of Montijo, June 28, 1805, 1256, RAM.

46. Report from D. Antonio Anento, doctor in the foundling house, about the trials carried out in 1805 and 1812, October 27, 1814, Correspondencia con médicos relativa a la Alimentación Artificial, Vacunas, etc, caja (box) 8342/7, ARM: "siguiendo este plan hasta el día 15 de noviembre." I will refer to these files from here on as "Correspondencia con medicos"; caja 8342/7 contains the correspondence of the Junta with the doctors they worked with.

47. Report from D. Antonio Anento: "Destinando para este efecto niños de toda edad, a saber, desde bautizos hasta los 18 meses en que se les desteta, mamando unos a la misma teta de la cabra tres mamadas, otros cuatro, algunos seis, ocho y más según la edad y robustez del niño."

48. Letter from the Countess of Montijo to Ignacio Ruiz de Luzuriaga, June 28, 1805, 1256, RAM: "Entre las varias disposiciones que ha tomado la Junta de Señoras . . . ha sido la de hacer una prueba de criar a los niños expósitos unos con Amas y otros con cabras, según propone en el Plan que establece el Médico Santiago García en la obra presentada a la Junta sobre la crianza física de los niños expósito que aprobó la Real Academia."

49. García, *Instituciones*, 221–30.

50. Report from D. Antonio Anento: "En que resolvió la Junta que algunas cabras se alimentaran con pienso de cebada, paja y algarrobas y otras continuaran pastando en el Campo, pero por no haber experimentado los efectos favorables que ansiosamente se esperaban, se cesó en estos ensayos, habiendo pasado a ejecutar iguales tentativas con la leche mamada de la misma teta de la Burra . . . observando todo cuanto ocurría en los Niños en cada una de ellas, para lo cual se formaba un diario legítimo de todas las novedades o síntomas que se advertían, prósperos o adversos. Hubo niños que mamaron de las cabras sin mezcla alguna de otros alimentos, otros que tomaron la teta de la cabra dos o tres veces al día, y lo restante según lo exigía la necesidad."

51. Report from D. Antonio Anento: "El 12 de Marzo se hizo disección anatómica en un niño llamado Eugenio, y en el estómago se le encontró un fuerte coágulo de leche."

52. Letter from the Countess of Montijo to Ignacio Ruiz de Luzuriaga, June 28, 1805, 1256, RAM.

53. Luis Granjel, *Historia de la Real Academia Nacional de Medicina* (Madrid: Taravilla, 2006).

54. José Ramón Bertomeu-Sánchez, "The Colic of Madrid (1788–1814): Experts, Poisons, Politics, and War at the End of the Ancien Régime in Spain," *Social History of Medicine* 33, no. 3 (August 2020): 728–48; Elena Serrano, "Spreading the Revolution: Guyton's Fumigating Machine in Spain. Politics, Technology, and Material Culture (1796–1808)," in *Compound Histories: Materials, Chemical Governance and Production*, ed. Lissa Roberts and Simon Werrett (Leiden: Brill, 2017), 106–30.

55. On Pedro Gutiérrez Bueno, see: José Ramón Bertomeu Sánchez, "Pedro Gutierrez Bueno (1745–1822) y las relaciones entre la quimica y la farmacia durante el ultimo tercio del siglo XVIII," *Hispania* 61 (2010): 539–61 and Antonio García Belmar and José Ramón Bertomeu Sánchez, "Pedro Gutiérrez Bueno (1745–1822), los libros de texto y los nuevos públicos de la química en el último tercio del siglo XVIII," *Dynamis: Acta Hispanica ad Medicinae Sientiarumque Historiam Illustradam* 21 (2001): 351–74.

56. On the origins, goals, members, and activities of this society, see Elena Serrano, "Sex and Prisons: Women and Spanish Penitentiary Reform, 1787–1808," in "The Sexes and the Sciences," ed. D. N. Wagner and J. Wharton, special issue, *Journal for Eighteenth-Century Studies* 42, no. 4 (December 2019): 501–17.

57. See for instance, letter from the Junta to the Madrid Economic Society, leg. 8421, Madrid, August 2, SEM, in which the Junta asked the society to fund Santiago García's translation of an English medical dictionary.

58. "Memoria Anual de la Junta" of 1805, 196/19, SEM: "El exactísimo diario de estos ensayos es el mejor testimonio de la prolijidad y esmero con que se ha procedido, pero no

habiéndose atrevido las Excelentísimas Socias a calificar por las resultas, se pasó este diario a la Real Academia Médica, cuyo parecer se espera." It was read at the meeting of the Madrid Economic Society on November 30, 1805.

59. Report from D. Antonio Anento: "El 15 de Noviembre resolvió la Junta que algunas cabras se alimentaran con pienso o cebada, paja y algarrobas, y otras continuaran pastando en el campo."

60. Note on the expenses incurred by the goats and the donkey in the years 1805 and 1806, Correspondencia con medicos, caja 8342/7, ARM. There is a note below: "The donkey began in December 1805. The trials were carried out from then until May 1806." See also "Memoria Anual de la Junta" of 1806, 200/17, SEM.

61. Letter from Rosario Cepeda to the Royal Academy of Medicine, March 29, 1806, 1330, RAM: "En la sesión que se celebró ayer hice presente el oficio de VS con el informe que acompaña de la comisión nombrada por esa Real Academia acerca del que en 28 de Junio del año pasado se la pidió sobre la crianza con cabras de los niños expósitos de la Real Casa de la Inclusa."

62. Report of the doctors of the Real Academy of Medicine of Madrid, 699 RAM: "La noticia, cierta o infundada, de que la Real Junta de Señoras de honor y mérito, bajo cuya dirección está aquel establecimiento, entorpeció desde luego los primeros pasos de la comisión." I refer to this hereafter as "La Comisión."

63. La Comisión, 699, RAM: "La Real Junta de Señoras de honor reuniendo a la dulzura de su sexo la ilustración propia de su clase y a sus talentos grandiosos una generosidad sin límites ha transformado aquella mansión en un albergue de beneficencia donde cada señora es una madre solícita y tierna, cada expósito un hijo querido objeto único de todos sus cariños." The use of "clase" here is ambiguous; I am inclined to think that it referred not to the social class but to the women belonging to the meritocracy.

64. La Comisión, 699, RAM: "La Junta ha excusado a la Real Academia de hacer este trabajo, administrando en su lugar leche de burra . . . Así el ingenio ayudado de la virtud alcanzó tanto como el arte y así las señoras de honor, dando una prueba más de su ilustración y su celo han establecido una regla inviolable en un asunto de incertidumbre."

65. See the report of the following year, "Informe annual de 1806," 200/17, SEM.

66. Report from D. Antonio Anento: "debo confesar que yo mismo estoy admirado de ver y haber tocado los efectos nada favorables que se experimentaron, y a penas hubo un niño que se pudiese conservar, ni mamando mucho, ni poco, con método, sin él, de la misma teta de la Cabra, como con pistero, aguada y pura."

67. Report from D. Antonio Anento: "Iguales sucesos se notaron en otros ensayos, que se ejecutaron en el mes de Abril de 1812 y terminaron en Mayo."

68. Letter from Ignacio Jáuregui to the Junta de damas, Correspondencia con medicos, caja 8342/7, ARM.

69. Letter from Ignacio Jáuregui to the Junta de Damas: "Este precioso manjar [la leche] tiene ya un principio de animalización y una tendencia a la iniciativa de la vida animal humana, de manera que resiste menos que toda otra sustancia a tomar la forma y cualidades de los humores propios de nuestra especie, y sugetarse a las leyes de la vida de asimilación."

70. Maria Rafaela Gutiérrez de Terán y González Vertiz (México,1771–?), wife of the second Count of de Casa Flores, ambassador of Spain in Vienna, Lisbon, and Saint Petersburg.

71. Dossier on the *nuevo sagú*, Correspondencia con medicos, caja 8342/7, ARM.

72. Draft of a letter from the Junta to the Madrid Economic Society, September 14, 1817,

ARM, caja 8342/7: "Cuyo experimento se ha empezado a hacer en la real casa de la inclusa bajo la dirección de la Exc.ª Señora."

73. Thanks to the anonymous reviewer for suggesting the second possibility.

74. Letter from Pascual Mora to the Junta. Results of the experiments with the flour of *nuevo sagú*, September 20, 1817, Correspondencia con medicos, caja 8342/7, ARM: "Que de ningún modo puede sustituir a las leches en los recién nacidos, porque necesitan de un alimento animalizado . . . que pueda penetrar por sus finísimos vasos lácticos."

75. Letter from Pascual Mora to the Junta, September 20, 1817: "No obstante, creo firmemente que la harina del nuevo Sagú . . . es preferible en ciertas circunstancias a otras clases de alimentos y aun a leche misma, pues si en uno la leche de cabras produce diarreas, y excita en las criaturas erupciones en la piel que tienen todo el carácter de depurativas, del mismo modo el nuevo Sagú corrige las diarreas."

76. Letter from Pascual Mora to the la Junta, September 27, 1817, Correspondencia con medicos, caja 8342/7, ARM: "El médico de la Inclusa da parte de la continuación del experimento"; letter from Pascual Mora to the Junta, October 3, 1817, caja 8342/7, ARM: "Nada se puede decir."

77. Report written by Andrés Alcón, who gives a positive (or favorable) assessment of the report written by Pascual Mora on the observations about the flour of the Maranta Arundinácea (*nuevo sagú*) carried out in the Madrid foundling house, October 23, 1820, 030 (1781), RAM.

78. Mora, *El hombre*. The supplement was called "Suplemento en el que se hace la debida apología de las leches en comparación con el método artificial y sus diferentes ensayos, con una noticia de las plantas exóticas e indígenas, de las que se extrae la fécula o harina que se ha querido substituir y aún preferir al orden de la naturaleza, y utilidades que les producen a los indios y americanos varios de sus vegetales." See *El Hombre*, 1:239–41.

79. "Instrucción para la excelentísima señora de Condesa de Casa Flores," in Mora, *El hombre*, 1:242n2.

80. Letter from the Countess of Alagón to the Countess of Casa-Sarrià, Correspondencia con medicos, caja 8342/7, ARM: "como sabes son 150 [criaturas] teniendo cada ama a 3, lo que además de ser perjudicial para los niños es perjudicial para las Amas." The Countess of Alagón, Maria del Pilar Silva y Palafox, Condesa de Castel-Florido, y Condesa viuda de Aranda (?–1835), became member of the Junta in 1799 and was president from April 1826 until 1828. In José Varela de Limia y Menéndez, *Los directores de la Real Sociedad Económica Matritense de Amigos del País y las presidentas de su Junta de Damas de Honor y Mérito* (Madrid: Talleres de "El Eco Franciscano," 1925). According to Limia y Menéndez, it was proposed that she revise the translation produced by the Marquise of Espeja of Erasmus Darwin's *A Plan for the Conduct of Female Education, in Boarding Schools, Private Families, and Public Seminaries.* She also authored the *Memoria expositiva de la situación de la Inclusa y Colegio de niñas de la Paz* (Madrid: Imprenta Eusebio Àlvarez, 1827).

81. Letter from the Countess of Alagón to the Countess of Casa-Sarrià, "A las Exma. Señora Condesa de Casa Sarrià de su compañera [Countess of Alagón]," in Correspondencia con medicos, caja 8342/7, ARM: "En este aspecto, aunque el médico ha separado los que le han parecido para criarlos con la leche de cabras, nada se adelanta, pues tenemos la experiencia de que es alivio momentáneo."

82. Letter from the Countess of Alagón to the Duchess of Benavente, July 18, 1828, Correspondencia con médicos, caja 8342/7, ARM: "por hallarme muda no voy en persona a pedír-

telos, pero siempre está pronta a complacerte tu amiga." And "Y como tu tuviste la curiosidad de llevar los apuntes de todo lo que se hacía y de las novedades que se notaban en las criaturas te estimaríamos mi compañera y yo si nos quisieras hacer el gusto de dejarnos tus apuntes que los copiaríamos y volveríamos al instante pues nos serían muy útiles para la nueva prueba."

83. Letter from the Duchess de Benavente to the Countess of Alagón, July 19, 1828, Correspondencia con medicos, caja 8342/7, ARM: "Mi querida amiga. Sabes puedes disponer de mi como gustes y lo que siento es que no se hallen entre mis papeles los que deseas, y te remito el único que por incidencia se toca algo sobre lactancia de cabras. Lo que tengo presente es que fue preciso abandonar este método, pues a pesar de la vigilancia de las Señoras se desgraciaron todos los niños. La falta de pastos y tener que alimentar las cabras a pienso engrosaban la leche, y no la digerían y se desvanecieron las esperanzas de que aquí se lograsen los felices resultados que en los Países extranjeros."

84. Letter from the Countess of Alagón to Countess of Casa-Sarrià (no date, no number) and "A las Exma, Sera Condesa de Casa Sarrià de su compañera," Correspondencia con medicos, caja 8324/7, ARM.

85. Demerson, *María Francisca*, 236: "En razón de los perjuicios e inconvenientes que podrían resultar a los niños de un ensayo que hasta ahora no tiene ninguna probabilidad en su favor."

Chapter 4. Learned Daughters

1. The Duchess of Benavente is also known as the Duchess of Osuna because of her marriage. In 1771 María Josefa de la Soledad Alonso Pimentel (1752–1834), Duchess of Benavente, married her cousin Pedro de Alcántara Téllez-Girón (1755–1807), the future ninth Duke of Osuna. To recall, the Junta de damas was the female branch of the Madrid Economic Society (see previous chapters of this book).

2. The children were: Josefa Manuela Téllez-Girón, the future Marquise of Camarasa (1783–1817); Joaquina Téllez-Girón, the future Marquise de Santa Cruz (1784–1851); Francisco de Borja Téllez-Girón, future tenth Duke of Osuna (1785–1820); and Pedro de Alcántara Téllez-Girón, Prince of Anglona (1786–1851). See also Juan J. Luna, "Retrato de los duques de Osuna y sus hijos," in *Realidad e imagen: Goya, 1746–1828*, ed. F. Torralba (Zaragoza: Gobierno de Zaragoza, 1996), 100; A. Reuter, "Family of the Duke and Duchess of Osuna, 1787–1788," in *Goya: Images of Women*, ed. J. A. Tomlinson (Washington, DC: National Gallery of Art, 2002), 162–64.

3. M. Moreno de las Heras, "The Family of the Duques de Osuna," in *Goya and the Spirit of Enlightenment*, ed. Alfonso E. Pérez Sánchez and Eleanor A. Sayre (Boston: Museum of Fine Arts, 1988), 38–40n17.

4. Mónica Bolufer Peruga, "Gender and the Reasoning Mind: Introduction," in *Women, Gender and the Enlightenment*, ed. Sarah Knott and Barbara Taylor (Basingstoke: Palgrave Macmillan, 2005): 189–94, here, 189; Gloria Franco Rubio, "El talento no tiene sexo: Debates sobre la educación femenina en la España moderna," in *El alma de las mujeres: Ámbitos de espiritualidad femenina en la modernidad (siglos XVI–XVIII)*, ed. Javier Burrieza Sánchez (Valladolid: Universidad de Valladolid, 2015), 363–93; Sally Ann Kitts, *The Debate on the Nature, Role and Influence of Woman in Eighteenth-Century Spain* (Lewiston, NY: Edwin Mellen Press, 1995).

5. This was for instance advised by José Isidoro Morales, *Comentario de D. Joseph Isidoro Morales, al Excmo. Señor D. Joseph de Mazarredo sobre la enseñanza de su hija* (Madrid: Gabriel

de Sancha, 1796), 43. But others, such as Rita Caveda, enthusiastically advised women to also study philosophy: Rita Caveda, *Cartas selectas de una señora a una sobrina suya, entresacadas de una obra inglesa* (Madrid: Garcia y Compañía, 1800), map VI: "Sobre las cualidades que deben tener las mujeres." A study of the *Cartas selectas* is available in Mónica Bolufer et al., *Mujeres y modernización: Estrategias culturales y prácticas sociales (siglos XVII–XX)* (Madrid: Instituto de la Mujer, 2008), 144–58.

6. Esperanza Mó Romero and Margarita Eva Rodríguez García, "Educar: ¿A quién y para qué?" in *Historia de las mujeres en España y América Latina*, ed. Isabel Morant Deusa (Madrid: Cátedra, 2005), 729–56, here 734.

7. Margarita Ortega, "La educación de la mujer en la ilustración española," in *Revista de la Educación* (1988): 303–26; Catherine M. Jaffe, "Contesting the Grounds for Feminism in the Hispanic Eighteenth Century: The Enlightenment and Its Legacy," in *The Routledge Companion to the Hispanic Enlightenment*, ed. Elizabeth Franklin Lewis, Mónica Bolufer Peruga, and Catherine M. Jaffe (London: Routledge, 2020), 69–82. On the importance of education in the French revolutionary ideology, see, for instance, Penny Brown, *A Critical History of French Children's Literature (1600–1830)* (London: Routledge, 2008), 189–218. On education as a mark of social identity, see Aileen Fyfe, "Reading Children's Books in Late Eighteenth-Century Dissenting Families," *Historical Journal* 43, no. 2 (June 2000): 453–73; and Neil McKendrick, John Brewer, and J. H. Plumb, *The Birth of a Consumer Society: The Commercialization of Eighteenth-Century England* (Bloomington: Indiana University Press, 1982), 286–315.

8. Catherine M. Jaffe, "Doña Leonara's Library: Women's Readings from *The Spectator* (1711) to *El Semanario de Salamanca* (1795)," in *Eve's Enlightenment: Women's Experience in Spain and Spanish America, 1726–1839*, ed. Catherine M. Jaffe and Elizabeth Franklin Lewis (Baton Rouge: Louisiana State University Press, 2009), 178–96, here 179; Jacques Soubeyroux, "Niveles de alfabetización en la España del siglo XVIII: Primeros resultados de una encuesta en curso," *Revista de Historia Moderna: Anales de la Universidad de Alicante* 5 (1985): 159–72.

9. See Carla Hesse, "Women Intellectuals in the Enlightened Republic of Letters: Introduction," in Knott and Taylor, *Women, Gender and the Enlightenment*, 259–63; Pilar Pérez Cantó and Rocío de la Nogal, "Las mujeres en la arena pública," in Deusa, *Historia de las mujeres*, 757–89, especially for the colonial context; on aristocratic women and sociability, see Theresa Ann Smith, *The Emerging Female Citizen: Gender and Enlightenment in Spain* (Berkeley: University of California Press, 2006), 41–50; and María Victoria López-Cordón Cortezo, "The Enlightenment and Its Interpreters: Nobility, Bureaucrats, and Publicists," in Franklin Lewis, Bolufer Peruga, and Jaffe, *Routledge Companion*, 203–17 and 206–7.

10. Clarissa Campbell Orr, "Aristocratic Feminism, the Learned Governess, and the Republic of Letters," in Knott and Taylor, *Women, Gender and Enlightenment*, 306–25. On women courtiers, see Nadine Akkerman and Brigit Houben, eds. *The Politics of Female Households: Ladies-In-Waiting across Early Modern Europe* (Leiden: Brill, 2014); Florian Kühnel, "'Minister-Like Cleverness, Understanding, and Influence of Affairs': Ambassadresses in Everyday Business and Courtly Ceremonies at the Turn of the Eighteenth Century," in *Practices of Diplomacy in Early Modern World, c. 1410–1800*, ed. Tracey A. Sowbery and Jan Hennings (London: Routledge, 2017): 130–42; Glenda Sluga and Carolyn James, eds., *Women, Diplomacy and International Politics since 1500* (New York: Routledge, 2016).

11. Paula Findlen, "Women on the Verge of Science: Aristocratic Women and Knowledge in Early Eighteenth-Century Italy," in Knott and Taylor, *Women, Gender and Enlightenment*, 265–87; Massimo Mazzotti, *The World of Maria Gaetana Agnesi, Mathematician of God* (Balti-

more: Johns Hopkins University Press, 2007). Mónica Bolufer Peruga, "Women of Letters in Eighteenth-Century Spain: Between Tradition and Modernity," in Jaffe and Franklin Lewis, *Eve's Enlightenment*, 17–32.

12. Mónica Bolufer and Elena Serrano, "Maritime Crossroads: The Knowledge Pursuits of María de Betancourt (Tenerife, 1758–1824) and Joana de Vigo (Menorca, 1779–1855)," in "Enlightened Female Networks," ed. Anna Maerker, Elena Serrano, and Simon Werrett, special issue, *Notes and Records: The Royal Society Journal of the History of Science* (forthcoming).

13. For a brief overview of classical children's book in the British and North America context, beginning in 1475, see Eric J. Johnson, "Chronology," in *The Cambridge Companion to Children's Literature*, ed. M. O. Grenby and Andrea Immel (Cambridge, UK: Cambridge University Press, 2009), xiii–xvii.

14. An excellent summary of the historiographic debates is found in M. O. Grenby, "The Origins of Children's Literature," in Grenby and Immel, *Cambridge Companion*, 3–18. See Andrea Immel, "Children's Books and Constructions of Childhood," in the same volume (19–34); Matthew O. Grenby, *The Child Reader, 1700–1840* (Cambridge, UK: Cambridge University Press, 2011); Andrew O'Malley, *The Making of the Modern Child: Children's Literature and Childhood in the Late Eighteenth Century* (New York: Routledge, 2003). On French juvenile literature, see Brown, *Critical History*, especially chs. 3 and 4; on Spanish children's literature, see Carmen Bravo-Villasante, *Historia de la literatura infantil española* (Madrid: Doncel, 1972); Jaime García Padrino, *Así pasaron muchos años: En torno a la literatura infantil española* (Cuenca: Ediciones de la Universidad de Castilla-La Mancha, 2001).

15. Mary V. Jackson, *Engines of Instruction, Mischief and Magic: Children's Literature in England from Its Beginning to 1839* (Lincoln: University of Nebraska Press, 1989); Brown, *Critical History*, 85–128; Julia Briggs, "'Delightful Task!': Women, Children and Reading in the Mid-Eighteenth Century," in *Culturing the Child 1690–1914: Essays in Memory of Mitzi Myers*, ed. Donelle Ruwe (Lanham, MD: Scarecrow Press, 2005), 67–84.

16. Briggs, "Delightful Task," 67–82. See also Ruth B. Bottigheimer, "The Book on the Bookseller's Shelf and the Book in the English Child's Hand," in Ruwe, *Culturing the Child*, 3–28.

17. François-Guillaume Ducray-Duminil, *Lolotte et Fanfan, ou les Aventures de deux enfans abandonnés dans une isle déserte, rédigées . . . sur des manuscrits anglais*, 2 vols. (Paris: Maradan, 1788). The book was translated into English as *Ambrose and Eleanor or the Adventures of Two Children Deserted on an Uninhabited Island* (London: R. & L. Peacock, 1796), while the Spanish translation was titled *Los dos Robinsones o aventuras de Carlos y Fanny*. I have consulted the second edition (Madrid: Justo de la Barra, 1797). Louis-François Jauffret, *Les Charmes de l'enfance et les plaisirs de l'amour maternel* (Paris: Moutard, 1791), in Spanish translated as *Las gracias de la niñez y placeres del amor maternal* (Madrid: Imprenta Real, 1804),

18. Louis-François Jauffret, *Les voyages de Rolando et de ses compagnons de fortune autour du monde* (Leipzig: J. C. D. Sinner, 1803); in Spanish, *Los viajes de Rolando y de sus compañeros de fortuna alrededor del mundo* (Madrid: Imprenta Real, 1804).

19. *The Children's Magazine; or, Monthly Repository of Instruction and Delight* (London: John Marshall, 1799); *The Young Gentleman's and Lady's Magazine* (London: J. Walker, 1799–1800), https://blogs.princeton.edu/cotsen/tag/young-gentlemans-and-ladys-magazine/.

20. Jean-Jacques Fillassier, *Eraste ou l'ami de la jeunesse* (Paris: Vincent, 1773).

21. James A. Secord, "Newton in the Nursery: Tom Telescope and the Philosophy of Tops and Balls (1761–1838)," *History of Science* 23, no. 2 (June 1985): 127–51. The title of the work

studied by Secord was *Tom Telescope: The Philosophy of Tops and Balls or the Newtonian System of Philosophy* (London: J. Newbery, 1761). For the major changes in the eighteenth-century and nineteenth-century editions of the *Philosophy of Tops and Balls,* see Secord, "Newton in the Nursery," appendix.

22. Jill Shefrin, "Make It a Pleasure and Not a Task: Educational Games for Children in Georgian England," *Princeton University Library Chronicle* 40, no. 2 (Winter 1999): 251–75; Brian Alderson, "New Playthings and Gigantick Histories: The Nonage of English Children's Books," *Princeton University Library Chronicle* 60, no. 2 (Winter 1999): 178–95.

23. On science resources for children in the late eighteenth- and nineteenth-century British context, see Melanie Keene, *Science in Wonderland: The Scientific Fairy Tales of Victorian Britain* (Oxford: Oxford University Press, 2015); idem., "From Candles to Cabinets: 'Familiar Chemistry' in Early Victorian Britain." *Ambix* 60 (2013): 54–77 and idem., "Playing among the Stars*: Science in Sport, or the Pleasures of Astronomy* (1804)." *History of Education: Sources and Interpretations* 40 (2011): 521–42.

24. Ann B. Shteir, "Priscilla Wakefield's Natural History Books," in *From Linnaeus to Darwin: Commentaries on the History of Biology and Geology,* ed. Alwayne Wheeler and James H. Price (London: Society for the History of Natural History, 1985), 29–36.

25. A useful summary can be found in Brown, *Critical History,* 157–88. See also: Adriana Benzaquen, "Childhood, Identity and Human Science in the Enlightenment," *History Workshop Journal* 57, no. 1 (Spring 2004): 35–57.

26. U. C. Knoepflmacher, "Children's Texts and the Grown-Up Reader," in Grenby and Immel, *Cambridge Companion,* 159–73, here 159.

27. Kimberley Reynolds, "Changing Families in Children's Fiction," in Grenby and Immel, *Cambridge Companion,* 193–208, especially 193–95; Jacqueline Pearson, *Women's Reading in Britain, 1750–1835* (Cambridge, UK: Cambridge University Press, 1999), 65–68.

28. Michel Manson, *Les livres pour l'enfance et la jeunesse sous la Révolution* (Montreuil: INRP, Bibliothèque Robert-Desnos, 1988), 20–23; Brown, *Critical History,* especially ch. 6; Isabelle Havelange and Ségolène Le Men, *Le Magasin des enfants: La littérature pour la jeunesse (1750–1830)* (Montreuil: Bibliothèque Robert Desnos, 1988); Jean Glénisson et Ségolène Le Men, ed., *Le Livre d'enfance et de jeunesse en France* (Bordeaux: Société des Bibliophiles de Guyenne, 1994); Jean Glénisson, "Le livre pour la jeunesse," in *Histoire de l'edition française,* ed. Henri-Jean Martin and Roger Chartier, 4 vols. (Paris: Promodis, 1986), vol. 3.

29. The second journal for children was called *Minerva de la Juventud española* (Madrid: Publishing house unknown, 1833–1834). See: Antonio Viñao Frago, "El libro escolar," in *Historia de la edición en España 1836–1936,* ed. Jesús A. Martínez Martín (Madrid: Marcial Pons, 2001), 309–36, here 318; Raquel Sánchez García, "La edición de libros infantiles y juveniles," in Martin, *Historia de la edición,* 337–54, here 340. See also Mercedes Chivelet, *La prensa infantil en España, desde el siglo XVIII hasta nuestros días* (Madrid: Editorial SM, 2009).

30. Paula Demerson, *Esbozo de Biblioteca de la Juventud Ilustrada (1740–1808)* (Oviedo: Cátedra Feijoo, 1976), 22. Demerson has published edited editions of the *La Gazeta de Madrid,* the *Memorial Literario,* and *El Revisor General.* See also Álvaro Ruíz de la Peña Soler, "Ilustración, lectura y juventud en la España del XVIII," in *Historia de la edición y la lectura en España 1472–1914,* edited by Victor Infantes, François López, and Jean-François Botrel (Madrid: Fundación Germán Sánchez Ruipérez, 2003), 481–98.

31. Norma Clarke, "The Cursed Barbault Crew: Women Writers and Writing for Children in the Late Eighteenth Century," in *Opening the Nursery Door: Reading, Writing and Child-*

hood (1600–1900), ed. Mary Hilton, Morang Styles, and Victor Watson (London: Routledge, 1997), 91–103. See also Pearson, *Women's Reading in Britain*, 65–68. An updated bibliography is found in Joanna Wharton, *Material Enlightenment: Women Writers and the Science of Mind, 1770–1830*, vol. 1 (Boydell & Brewer, 2018). See also Aileen Fyfe, "Introduction," *Science for Children*, ed. Aileen Fyfe, 7 vols. (Bristol: Thoemmes Press, 2003), 1:xi–xxii.

32. Emilio Palacios, *La mujer y las letras en la España del siglo XVIII* (Madrid: Ediciones Laberinto, 2002). Palacios pointed out that we know of around two hundred literary women, but in most cases only by name and a title, because most of the texts are lost. See also Maria del Pilar Zorrozua, *Escritoras de la Ilustración española (1759–1808)* (Bilbao: Dpto. Publicaciones Universidad Deusto, 1999).

33. Joachim Heinrich Campe, *El nuevo Robinson, historia moral reducida a diálogos para la instrucción y entretenimiento de niños y jóvenes de ambos sexos*, trans. Tomás de Iriarte (Madrid: Benito Cano, 1789). Tomás de Iriarte also wrote a manual on history and geography: *Lecciones instructivas sobre la historia y la geografía: Obra póstuma*, 3 vols. (Madrid: Imprenta Real, 1794). Some examples of what was published may be found in Alejandro Mayordomo Pérez and Luis Miguel Lázaro Lorente, *Escritos pedagógicos de la Ilustración* (Madrid: Centro de Publicaciones, MEC, 1989).

34. Valentín de Foronda, *Lecciones ligeras de química* (Madrid: Imprenta de Gonzalez, 1791).

35. Jean François Dubroca, *Conversaciones de un padre con sus hijos sobre la historia natural*, trans. Manuel María de Ascargorta (Madrid: Imprenta Real, 1802–1803). José Viera y Clavijo's known didactic poems were "Al globo aerostático," "Las cuatro partes del día," and "Los aires fijos"; the manual was called *Noticias del cielo o Astronomía para niños*. See Miguel Hernández González, *Opúsculos científicos de José Viera y Clavijo* (La Orotava: Fundación Canaria Orotava de Historia de la Ciencia, 2012).

36. Joaquín Álvarez Barrientos, "Cultura y Política entre siglos," in *Se hicieron literatos para ser políticos: cultura y política en la España de Carlos IV y Fernando VII*, ed. Joaquín Álvarez Barrientos (Cádiz: Servicio de Publicaciones de la Universidad de Cádiz, 2004), 11–24; Joaquín Álvarez Barrientos, "Los hombres de letras," in *La República de las letras en la España del siglo XVIII*, ed. Joaquín Álvarez Barrientos, François López, and Inmaculada Urzainqui (Madrid: Consejo Superior de Investigaciones Científicas, 1995), 19–62. On censorship and the Inquisition, see also, in the same volume, François López, "El libro y su mundo," 63–124.

37. Brown, *Critical History*, 191.

38. Jean-Luc Chappey, *La Société des observateurs de l'homme (1799–1804): Des anthropologues au temps de Bonaparte* (Paris: Société des études Robespierristes, 2002); Louis-François Jauffret, *Les merveilles du corps humain* (Paris: A. J. Dugour et Durand, 1798); idem., *Jeux zoologiques et géographiques* (Paris: Leclère, 1799); idem., *Éléments de zoographie, ou l'histoire des animaux considérés relativement au degré d'étendue des régions que chaque espèce occupe sur la surface du globe* (Paris: Demoraine, n.d.).

39. Pedro Ruiz Torres, "Reformismo e ilustración," in *Historia de España*, ed. Josep Fontana and Ramón Villares (Madrid: Crítica/Marcial Pons, 2007), 5:610–15; Francisco Comín, "Canga Argüelles: un planteamiento realista de la hacienda liberal," in *Economía y economistas españoles*, vol. 4, *La economía clásica*, ed. Enrique Fuentes Quintana (Barcelona: Galaxia Gutenberg, 2000), 413–39.

40. Anacreonte, *Obras de Anacreónte*, trans. Joseph and Bernabé Canga Argüelles (Madrid: Imprenta de Sancha, 1795); Píndaro, *Obras de Píndaro*, trans. Joseph and D. Bernabé Canga Argüelles (Madrid: Imprenta de Sancha, 1798); Pedro Rodríguez de Campomanes, *Discurso*

sobre el fomento de la industria popular (Madrid: Sancha, 1774); Francisco Martínez de la Mata, *Suplemento al apendice de la educacion popular: Contiene dos discursos de Francisco de la Mata . . . de la Orden Tercera de la Penitencia lo publica con algunas notas Don Josef Antonio Canga Argüelles y Cifuentes Prada Martínez de la Mata, Francisco* (Madrid: Imprenta de Sancha, 1794). For a discussion of the importance of Campomanes in Spanish economic policies, see chapter 1 of this book.

41. The Spanish title was "Oficial de la Real Caja de Amortización de Vales Reales dentro de la Secretaría de Hacienda." For an account of José Canga Argüelles in the history of public finances, see Joseph Fontana, *La Hacienda en la historia de España 1700–1931* (Madrid: Instituto de Estudios Fiscales, 1980), 80. The most well-known works of Joseph Canga Argüelles are *Elementos de la ciencia de la Hacienda* (London: Imprenta de A. Macintosh, 1825) and *Diccionario de Hacienda* (London: Imprenta española de M. Calero, 1826–1827).

42. A bibliography of his publications can be found in Palmira Fonseca, *Un hacendista asturiano: José Canga Argüelles* (Oviedo: Real Instituto de Estudios Asturianos, 1995). For his English period, see Vicente Llorens, *Liberales y Románticos: Una emigración española en Inglaterra* (1823–1834) (Madrid: Editorial Castalia, 1968). In England, he published two journals: *Ocios de españoles emigrados* (1824–1827) and *El emigrado observador* (1828–1829).

43. Bernabé would publish in the journal *Anales de historia natural* (1799), which later changed its name to *Anales de ciencias naturales* (1801–1804).

44. Juan Sempere y Guarinos (1754–1830) quoted Antonio Sancha (1720–1790), father of Gabriel, as an example of Spanish entrepreneurism. On the books printed by Sancha, see *Catálogo de los libros impresos por D. Antonio de Sancha*, ([Madrid?]: Publishing house unknown, [between 1772 and 1800]); on Sancha's life, see Calcografía Nacional, *Antonio de Sancha, reinventor de lecturas y hacedor de libros (1720–1790)* (Madrid: Real Academia de Bellas Artes de San Fernando, Calcografía Nacional, 1997); Emilio Cotarelo y Mori, *Biografía de D. Antonio de Sancha: Un gran editor español del siglo XVIII* (Madrid: Cámaras Oficiales del Libro de Madrid y Barcelona, 1924); Antonio Rodríguez-Moñino, *La imprenta de Don Antonio de Sancha (1771–1790)* (Madrid: Castalia, 1971). Antoine François Fourcroy, *Diario de los nuevos descubrimientos de todas las ciencias físicas, que tienen alguna relacion con las diferentes partes del arte de curar*, 2 vols. (Madrid: Imprenta de Sancha, 1792–1793). See also catalogue of 1797, *Catálogo de los libros impresos en Casa de Sancha* (Madrid, 1797 or later).

45. José Canga Argüelles, "Industrial, Depósito," in José Canga Argüelles, *Diccionario*, 3:304.

46. José Canga Argüelles, "Industrial, Depósito," *Diccionario*, 3:308: "Un cúmulo tan grande de productos, arreglados según este sistema, presentará de un golpe el grandioso cuadro de la industria de España y sus colonias, en todas las materias en que hasta ahora se han ejercitado los pueblos mas notables de la Europa: ofrecerá á primera vista los lugares vacíos que una industria sabiamente animada puede llenar dentro de poco tiempo; y nos dará por fin un régulo infalible para juzgar del estado en que se halla la industria de este pueblo, ser el primero en el orden de todas las potencias."

47. *Gazeta* I, 2: "La historia antigua nos ofrece ejemplos bien señalados del poder de la educación, que llegó a mudar el carácter de naciones enteras, bajo la vista misma de los filósofos que las reformaron."

48. *Gazeta* I, 21.

49. José Clavijo y Fajardo, *Prólogo a la traducción de la "Historia natural" del Conde de Buffon; estudio preliminar de José Luis Prieto Pérez* (La Orotava: Fundación Canaria Orotava de Historia de la Ciencia, 2001). See, in the same volume, Clavijo's letter to the Secretario de Estado

Francisco Saavedra from June 3, 1798, where he explained how the museum could serve to the advance the country: "Medios para hacer útil para la prosperidad de la nación española el R. Gabinete de Historia Natural." See also Oficio de Josef Clavijo y Fajardo dirigido al Secretario de Estado Francisco Saavedra, June 3, 1798, Archivo Museo Nacional Ciencias Naturales; and Antonio González, "El real Gabinete de Historia Natural," in *Madrid, ciencia y corte* (Madrid: Gráfica Futura, 1999), 247–51.

50. *Gazeta* I, 261.

51. All these quotations are from *Gazeta* II, 176, 271, 261.

52. Brown, *Critical History*, 89–93.

53. *Gazeta* I, 9–10: "En París se publica una obra muy semejante titulada *Correo de los niños*, por Jauffret y aunque tomaremos de él y de otros autores una gran parte de nuestro Periódico, sin embargo aun en lo que tomemos se harán varias ediciones y reformas que le constituirán una obra nueva, acomodada a nuestras costumbres, a la extensión de nuestra idea y a que los niños comprendan con la mayor facilidad las instrucciones que se les vayan comunicando."

54. Geoffrey V. Sutton, *Science for a Polite Society: Gender, Culture, and the Demonstration of Enlightenment* (Oxford: Westview, 1995).

55. Quoted from the edition edited by Louis-François Jauffret, *Voyage au jardin des Plantes, contenant la description des galeries d'histoire naturelle, des serres . . . de . . . l'école de botanique, avec l'histoire des . . . animaux de la ménagerie nationale* (Paris: C. Houel, 1797), 37: "Que font ici ces dames? Me dit-il; est-ce qu'elles veulent apprendre la chimie?"

56. Jauffret, *Voyage au jardin des Plantes*, 37–38: "Elles viennent apprendre de Fourcroy, que le nitre ou salpêtre s'appelle aujourd'hui nitrato de potasa, et que le sel de cuisine s'appelle muriate de soude; elles viennent apprendre à décomposer les élémens; à changer l'eau en air et l'air en eau; à connaitre les deux bases de l'air commun, qui sont l'oxygène et le gas azote."

57. On chemical itinerant lessons, see chapter 5 in this book. See also chapter 3 on the visit of the female society to the Royal Laboratory to measure the content of oxygen in the Madrid jails. On the scientific institutions at the time, see: Antonio Lafuente, *Guía del Madrid científico: ciencia y corte* (Madrid: CSIC, 1998); Antonio Lafuente and Juan Pimentel, "La construcción de un espacio público para la ciencia: Escrituras y escenarios en la Ilustración española," in *Historia de la ciencia y de la técnica en la Corona de Castilla*, ed. Luis García Ballester, 4 vols., (Valladolid: Junta de Castilla y León, 2002), 4:111–56; Antonio Lafuente and Nuria Valverde, *Los mundos de la ciencia en la Ilustración Española* (Madrid: Fundación Española para la Ciencia y la Tecnología, 2003).

58. *Gazeta* II, 35: "Pasábamos un día a hora de las once por la calle del Turco en compañía de nuestro papa y vimos varias personas que entraban a toda prisa por una de las muchas puertas que tiene la gran fábrica de los cristales. Preguntamos a Papá que a adonde iban y nos respondió que a las lecciones de química del señor Chavaneau [*sic*]."

59. From 1799, all three official chemistry laboratories in Madrid had been unified in the Escuela de Química; the director was another prestigious Frenchman, Joseph-Louis Proust (1754–1826).

60. *Gazeta* I, 8: "Los niños reciben la instrucción con más placer comunicada por la boca de otros niños."

61. Greg Myers, "Science for Women and Children: The Dialogue of Popular Science in the Nineteenth Century," in *Nature Transfigured: Science and Literature, 1700–1900*, ed. John Cristie and Sally Shuttleworth (Manchester: Manchester University Press, 1989), 170–200. See also chapters 1 and 5 of this book.

62. *Gazeta* I, 65: "Querida Luisita no puedo explicarte el alegre entusiasmo que causó en toda nuestra amable tertulia la lectura de tu respuesta a mi anterior: ¡cuántos elogios a tu inteligencia, cuántos tiernos conjuros por obligarme a continuar con mi empresa, y cuántos votos ardientes por volver a ver a la amable Luisita." *Gazeta* 2, 66: "Tu feliz comprensión manifestada en las bellísimas respuestas dadas a mis lecciones me llenan de lisonjeras esperanzas y me hacen tomar vanidad de tener una discípula tan respetable."

63. *Gazeta* II, 37: "Nos dijo que estar repartido por la atmósfera y por todos los cuerpos de nuestro planeta un fluido invisible llamado calórico. . . . Este fluido metiendo entre las partes de los cuerpos los hace mayores y más abultados," *Gazeta* II, 36: "Porque has de entender, amada Luisita, que todos los cuerpos de la naturaleza se componen de unas partecitas tan infinitamente pequeñas que no las alcanza a distinguir la vista ayudada de los mejores microscopios."

64. *Gazeta* II, 68–69.

65. *Gazeta* II, 97–106: "El pomito: diálogo en donde se continúan los principios de química."

66. *Gazeta*, II, 359: "Las substancias vegetales y animales en putrefacción producen una gran cantidad de los gases ácido carbónico, azoe e hidrógeno: y el agua, cuya descomposición produce también este último, suministra además el oxígeno, únicos elementos que entran en la composición de las especies vegetales. Así la tierra no obra en la vegetación sino en calidad de placenta, facilitando la descomposición de los abonos para reducir sus principios a gases."

67. *Gazeta* II, 129–42: "El jardín: diálogo donde se tratan varios puntos de geografía, botánica y artes."

68. Francesc Salvà i Campillo and Francesc Santapons, *Disertación sobre la explicación y uso de una Nueva Máquina para agramar cáñamos y linos inventada por los doc. en medicina* (Madrid: Imprenta Real, 1784).

69. *Gazeta* II, 257–71: "Las calles de Madrid: diálogo que contiene varios principios de economía pública y artes." Some examples are found on pages 265 and 270: "Para que abunden los géneros . . . basta que haya quien los compre: así estas naranjas vienen desde cuarenta y sesenta leguas a regalar el paladar de los madrileños sólo porque éstos las estiman y las pagan"; 270: "Aquí tenéis géneros de casi todos los puntos de la tierra, que el comercio trae a nuestras mismas casas, animado del premio o ganancia que saca por este trabajo." On the economic ideas of the Canga Argüelles, see Palmira Fonseca, *Un hacendista*, 308–10. Other readings of Canga Argüelles have been provided by James Mill, David Hume, Thomas Tooke, Antoine-Louis Destutt de Tracy, and Heinrich Storch.

70. *Gazeta* II, 271: "Todo esto amigos míos, debe haceros amar la sociedad y la unión de los hombres, respetar las artes, la agricultura y el comercio, únicas fuentes de todos los bienes que gozamos y trabajar con todas vuestras fuerzas por ser útiles en algo a los que tanto se fatigan en serlo para vosotros."

71. *Gazeta* II, 262–63.

72. In 1771 María Josefa de la Soledad Alonso Pimentel (1752–1834), Duchess of Benavente, married her cousin Pedro de Alcántara Téllez-Girón (1755–1807), the future Ninth Duke of Osuna.

73. The duchess asked D. Andrés Celle (tutor of the children of the Count of Fernán Nuñez), who sent her a report from Atanasio Puyal (who would become a bishop in 1815)—both of whom were teachers of Diego Clemencín in the seminar of Murcia; she also asked Alonso Camacho and the well-known publisher Joaquin Ibarra. Ibarra interviewed people who knew Clemencín well and wrote back to the duchess that Clemencín was of "kind character, outstanding talent, and assiduous in the study." His education comprised Latin and Greek, theology, and modern philosophy, which included metaphysic, logic, and theo-

retical and experimental physics. Ricardo Escavy Zamora, "Estudio introductorio," in *Diego Clemencín. Lecciones de Gramática y Ortografía Castellana: Edición Facsimilar de la edición de Madrid; Imprenta de D. Miguel de Burgos, 1842* (Murcia: Universidad de Murcia, 2012), 11–88, 18–22. A transcription of the reports can be found in Jean Sarrailh, "D. Diego Clemencín," *Bulletin Hispanique* 24 (1922): 125–30. On the extraordinary life of Clemencín, who became a famous commentator on *Don Quijote*, a member of the Royal Academy of History (1800) and the Royal Academy of Language (1805), and a member of the liberal party, see Antonio López Ruiz and Eusebio Aranda-Muñoz, *Don Diego Clemencín (1765–1834), Ensayo bio-bibliográfico* (Murcia: Publicaciones Universidad de Murcia, 1948).

74. Paloma Fernández-Quintanilla, *La IX Duquesa de Osuna: Una ilustrada en la Corte de Carlos III* (Madrid: Ediciones 12 Calles, 2017), 84–89.

75. "En consecuencia, desde esta misma hora, me tengo por admitido en su casa . . . persuadido a que nunca me sería tan ventajoso como servir a una Señora de tanta ilustración y de tan bello modo de pensar en orden a la educación de sus hijos." Translation taken from Moreno de las Heras, "Family of the Duques de Osuna," 38–40, here 40. Clemencín's letter is printed in its entirety in Escavy Zamora, "Estudio introductorio," 57.

76. *Examen público de las señoras Doña Josefa y Doña Joaquina Girón y Pimentel, y de los señores Francisco y D. Pedro sus hermanos hijos de los excelentísimos señores Duques de Osuna Condes-Duques de Benavente, el día 20 de abril de 1797* (Madrid: Benito Cano, 1797), 39. Clemencín also wrote a short manual on grammar and orthography in dialogue format, which was published posthumously, as well as two manuals on geography and natural history. Diego Clemencín, *Lecciones de gramática y ortografía castellana. Edición facsimilar de la edición de Madrid* (Madrid: Imprenta de D. Miguel de Burgos, 1842).

77. Diego Clemencín, *Discurso leido en la abertura del exâmen público de las señoras doña Josefa y doña Joaquina Giron y Pimentel, y de los señores d. Francisco y d. Pedro, sus hermanos, hijos de los excelentísimos señores Duques de Osuna, Condes-Duques de Benavente por don Diego Clemencín, el día 20 de abril de 1797* (Madrid: Benito Cano, [1794?]). See also: Diego Clemencín, *Proyecto para la Educación del Exmo. Sr. Marqués de Peñafiel, y del Sr. Príncipe de Anglona. Dirigido a sus padres, los Exmos. Duques de Osuna, Condes-Duques de Benavente.*

78. *Examen público*, 39: "Concluido el exámen, los señores don Francisco y don Pedro esperan del favor de los concurrentes se sirvan a pasar a su cuarto, donde mostrarán su gabinete de historia natural, especificando las piezas más notables que poseen de los tres reinos. Enseguida enseñarán su gabinete de máquinas e instrumentos de física y practicarán varios experimentos, especialmente de las máquinas neumáticas y eléctricas."

79. Virginia Albarrán, *El desafío del blanco: Goya y Esteve, retratistas de la casa de Osuna; a propósito de la donación de Alzaga* (Madrid: Museo Nacional del Prado, 2017), 21–27.

80. In 1790, for example, the duke and duchess paid the exorbitant price of 7,000 *reales de vellón* to the Spanish instrument maker Caledonio Rostriaga for the electric machine. The possession of a telescope, a microscope, a thermometer, a barometer, and a camera obscura was also documented.

81. *Project de Bibliothèque dressé d'après les notes remises par S.E. Madame la Duchesse d'Osuna [Manuscrito]*, S.XIX [I] Julian, Martín Abad, Mss. Bibli. BN, p. 107, n. 188, 88 p. See Julián Martín Abad, "Crecimiento de la colección de Manuscritos de la Biblioteca Nacional en el siglo XIX: Breves apuntes para una historia necesaria," *Boletín de la ANABAD* 42, no. 1 (1992).

82. See chapter 1 and 5 for more on these educational works.

83. López, "El libro y su mundo," 63–124, 85–93.

84. *Gazeta* II, 224: A. Berquin, *Introduction familière à la connoissance de la nature, traduction libre de l'anglois de Mrs. Trimmer* (Paris: Bureau de l'Ami des enfans, 1784); *Gazeta*, I, 192, *Diario de las madres de familia*; *Gazeta*, II, 128.

85. Mónica Bolufer Peruga, "Pedagogía y moral en el siglo de las luces: Las escritoras francesas y su recepción en España," *Revista de Historia Moderna. Anales de la Universidad de Alicante* 20 (2002): 5–65. Mónica Bolufer Peruga, "Conversations from a Distance: Spanish and French Eighteenth-Century Women Writers," in *A Companion to Spanish Women's Studies*, ed. Xon de Ros and Geraldine Hazbun (Woodbridge: Tamesis, 2011), 175–188.

86. The literature on this topic is vast. The following works offer various useful perspectives: Norma Clarke, *The Rise and Fall of the Woman of Letters* (London: Pimlico, 2004); Andrea Immel and Michael Witmore, "Little Differences: Children, Their Books, and Culture in the Study of Early Modern Europe," in *Childhood and Children's Books in Early Modern Europe, 1550–1880*, ed. Andrea Immel and Michael Witmore (New York: Routledge, 2006), 1–20, esp. 7; Hesse, "Women Intellectuals." See also Alan Rauch, "Mentoria: Women, Children and the Structures of Science," *Nineteenth-Century Contexts* 27, no. 4 (December 2005): 335–51; Isabel Morant, "Reasons for Education: New Echoes of the Polemic," in Jaffe and Franklin Lewis, *Eve's Enlightenment*, 51–61 and Alexandra Prunean, "Commercial Strategies in Paratextual Features of Late Eighteenth-Century Children's Books," PhD diss., Universidad Autónoma de Barcelona, 2016.

87. Jill Shefrin, "'Governesses to Their Children': Royal and Aristocratic Mothers Educating Daughters in the Reign of George III," in Immel and Witmore, *Childhood and Children's Books*, 181–211; Mitzi Myers, "Impeccable Governesses, Rational Dames, and Moral Mothers: Mary Wollstonecraft and the Female Tradition in Georgian Children's Books," *Children's Literature* 14 (1986): 31–59; William McCarthy, "Mother of All Discourses: Anna Barbault's Lessons for Children," in Ruwe, *Culturing the Child*, 85–111.

88. I cite the edition of 1842. Accessed October 6, 2020: https://books.google.com.ni/books?id=bt9QDITOJVQC&printsec=copyright#v=onepage&q&f=false. The first Spanish translation appeared in 1788. Stéphanie Félicité, Comtesse de Genlis, *Las veladas de la quinta o novelas e historias sumamente útiles para que las madres de familia, a quienes las dedica la autora, puedan instruir á sus hijos, juntando la doctrina con el recreo*, trans. Fernando de Guilleman, 3 vols. (Madrid: M. González, 1788). One of the exemplars in the Biblioteca Nacional de España (edition of 1804) belonged to the Duchess of Benavente.

89. "Y por esto su obra [por Genlis] es superior a la de cualquier hombre por sabio e instruido que sea; porque éste sólo escribe por especulación y aun cuando tenga alguna práctica nunca llega a lo que una Madre logra cuando ella misma educa a sus Hijos, mayormente si tiene talento y reflexión." Fernando Guilleman was a fellow of the Real Academia de Historia, where he collaborated on several articles (for example, about the uses of the barometer) and on the *Diccionario geográfico e histórico de España*. Antonio López Gómez and Carmen Manso, *Cartografía del siglo XVIII: Tomás López en la Real Academia de la Historia* (Madrid: Real Academia de la Historia, 2006).

90. Genlis, *Las veladas*, 3: "Dedico estas novelas con el mayor afecto y veneración a la Respetable Sociedad de Señoras unidas a la Sociedad Matritense, como representantes de todo el Cuerpo de Señoras del Reino."

91. Bolufer, "Pedagogía y moral"; Smith, *Emerging Female Citizen*; Elizabeth Franklin Lewis, *Women Writers in the Spanish Enlightenment: The Pursuit of Happiness* (Aldershot: Ashgate, 2004).

92. Libro de Actas de la Junta de damas, 1795–1796, Libro A/56-4, February 19, 1795, Archivo de la Sociedad Económica Matritense (Archivo Sociedad Económica Matritense, SEM): "La presidenta leyó un papel proponiendo a la Junta convendría se establecieran las dos comisiones de educación que previene el título 8° de los Estatutos": "Y porque debiendo en cualquier situación en que se encuentre, ya de Esposas, Madres o Amigas, tener tanto influjo en las operaciones de los hombres constituidos en estado civil y político, no podrían, a pesar de su natural dulzura, comunicándoles ideas útiles si se descuidara su educación."

93. Smith, *Emerging Female Citizen*.

Chapter 5. Imagining a New Country

1. Stéphanie Félicité, Comtesse de Genlis, *Les veillées du château* (Paris: M. Lambert et F.-J. Baudouin, 1782). In Spain, the book was translated as *Las veladas de la quinta*, 3 vols. (Madrid: Imprenta M. González, 1788).

2. Antoine Augustin Parmentier, *Bibliothèque universelle des dames: Économie rurale et domestique*, 8 vols. (Paris: Gaspard-Joseph Cuchet, 1788–1797).

3. "Memoria en forma de carta enviada de la Condesa de . . . a M. Dupuy sobre algunas bonificaciones en las tierras," in José Antonio Valcárcel, *Agricultura general y gobierno de la casa de campo*, 10 vols. (Valencia: Estevan Dolz, 1765–1795), 2:394–405.

4. Valcárcel, *Agricultura general*, 2:401: "Así, sin embargo de ser una mujer hecha a la gran Ciudad, vea V. cómo he sabido usar de las armas que me ha dado."

5. A good synthesis of the complex relation of the Enlightenment with rural topics inspired by the pioneer work of Roy Porter may be found in Simon Schaffer, "Enlightenment Brought Down to Earth," *History of Science* 41, no. 3 (September 2003): 257–68. See also Roy Porter, "The Environment and the Enlightenment: The English Experience," in *The Faces of Nature in Enlightenment Europe*, ed. Lorraine Daston and Gianna Pomata (Berlin, 2003), 17–38.

6. Denis Diderot, "Rustic Economy," *The Encyclopedia of Diderot & d'Alembert Collaborative Translation Project*, trans. Ann-Marie Thornton (Ann Arbor: Michigan Publishing, University of Michigan Library, 2013), http://hdl.handle.net/2027/spo.did2222.0001.953. Translation of "Economie rustique," *Encyclopédie ou Dictionnaire raisonné des sciences, des arts et des métiers*, vol. 5 (Paris: Briasson, 1755). On the concept of "economy" at the time, see Lissa Roberts, "Practicing Oeconomy during the Second Half of the Long Eighteenth Century: An Introduction," *History and Technology* 30, no. 3 (2014): 133–48; Emma Spary, *Feeding France: New Sciences of Food, 1760–1815* (Cambridge, UK: Cambridge University Press, 2014), 1–20; Simon Werrett, *Thrifty Science: Making the Most of Materials in the History of Experiments* (Chicago: University of Chicago Press, 2019), 15–41 and 177–86; and Margaret Schabas and Neil Di Marchi, "Introduction to Oeconomies in the Age of Newton," in "Oeconomies in the Age of Newton," ed. Schabas and Di Marchi, special issue, *History of Political Economy* 35 (2003) (suppl. 1): 1–13, here, 4.

7. See the list in Diderot's article on "Rural Economy" in the *Encyclopédie* and in *Semanario de agricultura y artes dirigido a los párrocos* vol. 1 (1799): 3–4: "Conocimientos que ha de tener un labrador instruido," which was an excerpt of the article "Agriculture" in the Abbé Rozier's *Course d'agriculture* (which was quickly translated into Spanish, in 1797).

8. Peter M. Jones, *Agricultural Enlightenment: Knowledge, Technology, and Nature 1750–1840* (Oxford: Oxford University Press, 2016); Charles C. Gillispie, *Science and Polity in France at the End of the Old Regime* (Princeton, NJ: Princeton University Press, 1980), 335–38; for Spain, see Lluís Argemí, "Agronomía y revolución agraria en España (1750–1820)," in *Agronomía y*

fisiocracia en España (1750–1820), ed. Ernest Lluch and Lluís Argemí (Valencia: Institución Alfonso El Magnánimo, 1985), 1–43; Rafael Serrano, "Técnicas agrícolas y zootecnia," in *Historia de la ciencia y de la técnica en la corona de Castilla*, ed. Luis García Ballester, 4 vols. (Madrid: Junta de Castilla y León, 2002), 4:603–30; Gonzalo Anes, "La agricultura española desde comienzos del siglo XIX hasta 1868: Algunos problemas," in *Ensayos sobre la economía española a mediados del siglo XIX*, ed. Pedro Schwartz and Gabriel Tortella Casares (Madrid: Ariel, 1970), 235–63; Jordi Cartaña i Pinén, "La agronomía en la España del setecientos," in *Técnica e ingeniería en España (III): El siglo de las luces; De la industria al ámbito agroforestal*, ed. Manuel Silva Suárez (Zaragoza: Prensas Universitarias de Zaragoza, 2005), 409–52.

9. On the question of the new experts in everyday practices, see especially Ursula Klein and Emma C. Spary, "Introduction: Why Materials?" in *Materials and Expertise in Early Modern Europe: Between Market and Laboratory*, ed. Ursula Klein and Emma C. Spary (Chicago: University of Chicago Press, 2009), 1–24. See also in the same volume Emma C. Spary, "Liqueurs and the Luxury Marketplace in Eighteenth-Century Paris," 225–57 and Barbara Orland, "Enlightened Milk: Reshaping a Bodily Substance into a Chemical Object," 163–97.

10. On Parmentier, see Alex Berman, "Parmentier, Antoine-Agustin," in *Dictionary of Scientific Biography*, ed. Charles C. Gillispie (New York: Charles Scribner's Sons, 1980); Joseph Antoine F. Balland, *La Chimie alimentaire dans l'oeuvre de Parmentier* (París, J.-B. Baillière et fils, 1902); and Brian Philip Block, "Liberté, égalité, Parmentier! A Critical Study of the Life and Work of Antoine-Augustin Parmentier, Pharmacist Extraordinaire" (PhD diss., Birkbeck College, University of London, 2009). On the politics of the food research carried out by Rumford and Parmentier, see Fritz Redlich, "Science and Charity: Count Rumford and His Followers," *International Review of Social History* 16, no. 2 (August 1971): 184–216; Rebecca Earle, "Food, Colonialism and the Quantum of Happiness," *History Workshop Journal* 84 (Autumn 2017): 170–93.

11. Charles Coulston Gillispie, ed., *Dictionary of Scientific Biography*, 18 vols. (New York: Charles Scribner Sons, 1983), 3:198. See also Jean-Antoine Chaptal, *De l'industrie française* (Paris: Antoine-Augustin Renouard, 1819), where he details many of these domestic procedures.

12. Gillispie, *Science and Polity in France*, chapter 2.

13. François Rozier, *Curso completo ó Diccionario Universal de Agricultura: Teórica, Práctica, Económica, y de Medicina rural y veterinaria . . .* , 16 vols. (Madrid: Imprenta Real, 1797–1803), 6:xxxii.

14. The bibliography on the role of locals, correspondents, and go-betweens is vast. For a survey, see Simon Schaffer et al., eds., *The Brokered World: Go-Betweens and Global Intelligence, 1770–1820* (Sagamore Beach: Science History, 2009), and Londa Schiebinger and Claudia Swan, *Colonial Botany: Science, Commerce and Politics in the Early Modern World* (Philadelphia: University of Pennsylvania Press, 2005). As an example of the botanical networks of Linnaeus, see Staffan Müller-Wille, "Reproducing Species," in Raymond Stephanson and Darren N. Wagner, *The Secrets of Generation: Reproduction in the Long Eighteenth Century* (Toronto: Toronto University Press, 2015), 37–58.

15. Jonathan R. Topham, "Historizing Popular Science," *Isis* 100, no. 2 (June 2009): 310–18; idem., "Publishing Popular Science," in *Science in the Marketplace: Nineteenth-Century Sites and Experiences*, ed. Aileen Fyfe and Bernard Lightman (Chicago: University of Chicago Press, 2007), 135–68; idem., "Rethinking the History of Science: Popularization/Popular Science," in *Popularizing Science and Technology in the European Periphery, 1800–2000*, ed. Faidra

Papanelopoulou, Agustí Nieto-Galan, and Enrique Perdiguero (Aldershot: Ashgate, 2009), 1–20. See also chapters 1 and 4 of this volume.

16. A discussion on women correspondents in other contemporary scientific networks can be found in Carl Thompson, "Women Travelers, Romantic-Era Science and the Banksian Empire," *Notes and Records: The Royal Society Journal of the History of Science* 73, no. 4 (December 2019): 431–55.

17. See the influential Londa Schiebinger, "Women of Natural Knowledge," in *The Cambridge History of Science*, vol. 3, ed. Katherine Park and Lorraine Daston (Cambridge, UK: Cambridge University Press, 2006), 192–205, here 194, and Londa Schiebinger, *The Mind Has No Sex? Women in the Origins of Modern Science* (Harvard University Press, 1989). See also the dissertation of Isabelle Lémonon Waxin, "La Savante des Lumières française, histoire d'une persona: pratiques, représentations, espaces et réseaux" (PhD diss., EHESS, 2019) and "Enlightened Female Networks: Gendered Ways of Producing Knowledge (1720–1830)," ed. Anna Maerker, Elena Serrano, and Simon Werrett, special issue, *Notes and Records: The Royal Society Journal of the History of Science* (forthcoming).

18. André J. Bourde, *Agronomie et agronomes en France au XVIIIe siècle* (S.E.V.P.E.N, 1967), 2:22. Arthur Young, *Travels in France during the Years 1787, 1788, 1789* (London: W. Richardson, Royal-Exchange, 1794), 1:180.

19. Charles Loïc and Christine Théré, "*Les femmes économistes*: The Place of Women in the Physiocratic Community," in Maerker, Serrano, and Werrett, "Enlightened Female Networks." These included the already mentioned Duchess d'Enville, Marie-Louise de Fourqueux (née Marie-Louise Auget de Montyon [1728–1798]), Marie de Malvieux (1725–after 1792), better known as Madame de Pailly, the Comtesse de Rochefort, Vicomtesse d'Aubusson, and the poet Anne-Marie du Boccage.

20. Tom Williamson, *Polite Landscapes: Gardens and Society in Eighteenth-Century England* (Stroud: Alan Sutton, 1995); Tim Richardson, *The Arcadian Friends: Inventing the English Landscape Garden* (London: Bantam, 2007); Terence M. Russell and Anne Marie Thornton, *Gardens and Landscapes in the* Encyclopédie *of Diderot and D'Alembert: The Letterpress Articles and Selected Engravings* (Aldershot: Ashgate, 1999).

21. Francesco Griselini, *Discurso sobre el problema de si corresponde a los párrocos y curas de las aldeas el instruir a los labradores en los buenos elementos de la Economía Campestre.* (Zaragoza: Blas Miedes, 1784); the Bishop of Barcelona, Pedro Díaz de Valdés, presented to the Royal Society of the Basque Provinces of Friends of the Country the memoire *El padre de su pueblo, o medios para hacer temporalmente felices a los pueblos, con el auxilio de los señores curas párrocos* in 1793, which was later republished: Pedro Díaz de Valdés, *Memoria premiada por la Real Sociedad Bascongada e impresa en Victoria en 1793. Reimpresa ahora, con un discurso previo y algunas notas* (Barcelona, Manuel Texero, 1806).

22. *Semanario*, "Prospecto," x: "Pero, ¿cuál será el medio de llevar a la noticia de nuestros labradores tan apreciable enseñanza cuando sabemos que en España los que labran no leen, y los que leen no labran?" A third editor, Domingo García Fernández, was fired shortly after being hired. Elisabel Larriba and Gérard Dufour, "Introducción," in *El Semanario de agricultura y artes dirigido a los párrocos (1797–1808)* (Valladolid: Editorial Ámbito, 1997), 9–62.

23. Emilio La Parra, *La alianza de Godoy con los revolucionarios: España y Francia a fines del siglo XVIII* (Madrid: CSIC, 1992); Richard Herr, *The Eighteenth-Century Revolution in Spain* (Princeton, NJ: Princeton University Press, 1958), especially 348–75.

24. The first two volumes of the *Encyclopédie méthodique* had already been published, while

the third was censored and prevented from being published by the Inquisition until 1792. See Herr, *Eighteenth-Century Revolution*, 349.

25. Herr, *Eighteenth-Century Revolution*, 348–75.

26. Jesús Pradells, "Juan Bautista Virio (1753–1837): Experiencia europea y reformismo económico en la España ilustrada," *Revista de historia moderna*, 8–9 (1988–1989), 233–71; Alberto Gil and Ana Boned, eds., *Diccionario biográfico del trienio liberal* (Madrid: El Museo Universal, 1991).

27. In his *Memoires*, Godoy praised Melón for his involvement in the extraction of sugar from beetroot. Melón was a close friend of the poet Leandro Fernández de Moratín (1760–1828), who quoted Melón many times in his diaries.

28. Elena Serrano, "Making Oeconomic People: The Spanish Magazine of Agriculture and Arts for Parish Rectors (1797–1808)," *History and Technology: An International Journal* 30, no. 3 (2014): 149–76.

29. Elisabel Larriba, "L'art au service de la divulgation scientifique: Le rôle des gravures dans le *Semanario de Agricultura y Artes dirigido a los Párrocos* (1797–1808)," in *El argonauta español* 2 (2005), http://argonauta.imageson.org/documents57.html.

30. Until 1805 the editorial offices were in Calle San José, now called Calle de Bárbara de Braganza. In 1805 the *Semanario* was relocated to the Royal Botanical Garden.

31. There is, for instance, an announcement that seeds will be shared in *Semanario*, October 22, 1801, no. 251, vol. X, 266, from the botanist D. Josef Pavón (from the botanical expedition to Peru), pertaining to seeds of the genus *Bombai* (the editors also advertise the sharing out of seeds from the Royal Gardens in Aranjuez).

32. *Semanario* III (1798), 151: "Entre las memorias de la Academia de Stockolmo, se lee lo que se ha de hacer con el lino para ponerlo tan hermoso como el algodón"; *Semanario* X (1801), 126: "Noticia del nuevo método de lavar la ropa de lino en las casas particulares leída por Chaptal"; *Semanario* X (1801), 173: "Modo de purificar la miel para que pueda servir en lugar de azúcar en los usos domésticos." A footnote explained that Gutiérrez Bueno tested the method in his laboratory.

33. Elaine Leong and Alisha Rankin, eds., *Secrets and Knowledge in Medicine and Science, 1500–1800* (Farnham: Ashgate, 2011).

34. *Semanario* X (1801), 232: "El álcali volátil es el que sostiene en el aire toda especie de miasmas nocivos"; Rozier's *Diccionario* I, 110: "Llamamos manteca a la parte crasa aceitosa e inflamable de la leche: se halla ésta distribuida y mezclada entre las partículas serosas y caseosas, de las cuales se va desprendiendo y subiendo a la superficie, en las vasijas en que se deja reposar la leche, y forma una tez espesa y fluida que se llama nata."

35. *Semanario*, "Prospecto," (1797), xii: "Se comuniquen de unas provincias a otras cuantos conocimientos convengan al adelantamiento de la agricultura y artes anexas, haciendo que los pueblos y los labradores no estén como aislados . . . la falta de esta comunicación es la causa de que en Catalunya, por ejemplo, se cultive una semilla importante que no se conoce en Sevilla."

36. *Semanario*, I (1797), xiv: "Este periódico será un centro de reunión de noticias provechosas a nuestros agricultores . . . excitará la curiosidad del hacendado y del artista industrioso a que hagan los experimentos que se les indiquen, y aun den parte al Semanario de sus resultados para publicarlos en beneficio común."

37. Serrano, "Making Oeconomic People," 7.

38. Pedro Rodríguez de Campomanes, *Discurso sobre el fomento de la industria popular* (Madrid: Imprenta de Sancha, 1774), cixv.

39. *Semanario* I (1797), 6. "Arte de hacer jabón en las fábricas y en las casas particulares." It began in the first issue and continued during several weeks.

40. *Semanario* I (1797), 253: "Todo el pueblo está contento con este nuevo invento, que ya se ha repetido en cuatro casas."

41. Elisabel Larriba, "Un intento de reforma agraria por y para las clases productoras: el Semanario de Agricultura y artes dirigido a los párrocos," *Brocar: Cuadernos de Investigación Histórica* 23 (1999): 87–117. See also idem., "Un instrument de la politique agraire de Godoy: Le Semanario de Agricultura y artes dirigido a los Párrocos," *Bulletin Hispanique* 104 (2002): 243–61; and Elisabel Larriba and Gérard Dufour, "Introducción," in Larriba and Dufour, *El Semanario de agricultura y artes*, 9–61.

42. Elisabel Larriba counts 186 different correspondents. Fifty-four signed as clergymen, twenty-one signed only with their initials, and fifty-six specified their labor activity. Some identified themselves as "farmers and cultivators" (15), other as landowners (7), botanists (2), members of the military (7), surgeons (3), doctors (2), apothecaries (2), mayors (3), attorneys (3), and entrepreneurs (*fabricantes*) (2). See Larriba, "Un intento de reforma agraria."

43. *Semanario* II (1797), 10: "Diálogo entre un párroco del Arzobispado de Toledo y un feligrés suyo sobre la utilidad del Semanario y perjuicios que sufren los labradores pobres."

44. AHN, Sección Consejos, 1893: "1796: Expediente formado en virtud de la Real Orden de su Majestad con que se ha remitido al Consejo un ejemplar del prospecto de un Semanario."

45. Larriba, "Un intento de reforma agraria," 87–117. Larriba detects a peak during the first part of 1798, where the number of letters reached around forty, almost 25 percent of the volume. See also idem., *Le public de la presse en Espagne à la fin du XVIIIe siècle (1781–1808)* (Paris: Honoré Champion, 1998).

46. *Semanario* III (1798), 182–86: "Carta de un subscriptor de Lérida."

47. *Semanario* III (1798), 183: "Con todo es digna de elogios por su industriosa invención, que yo he procurado esparcir por esta ciudad con algún provecho. ¡Ojalá se introdujera en lo demás de España! que de este modo no saldría tanto dinero del reino."

48. Roberts, "Practicing Oeconomy"; Werrett, *Thrifty Science*, 1–15; Serrano, "Making Oeconomic People," 16.

49. *Semanario* III (1798), 399–400: "Carta de una señora de Lucillos sobre el uso de las gachas."

50. *Semanario* IV (1798), 381–82: "Carta sobre las utilidades de una fábrica de paños de Aldea del Río."

51. *Semanario* II (1797), 318: "Belaunde (Doña María) sobre la fabricación del papel refundido."

52. María Agustina Romana de Siles y Cuenca accompanied her husband on his three diplomatic postings (Stockholm, The Hague, and London).

53. *Semanario* I (1797), 399: "Muy apreciable sería que los nacionales dispersos . . . como el celo tan plausible como el que distingue a la Señora que ha dirigido a los Editores la carta que precede."

54. *Semanario* IX (1801), 2–8, 6: "con la dulzura y afabilidad que exijo de ellos en el trato con aquellos honrados vecinos."

55. *Semanario* 1 (1797), 223–32; 240–48; 255–62.

56. Paula Carrasco, "Vida y obra de Pedro Gutiérrez Bueno," *Boletín de la Sociedad Española de Historia de la Farmacia* 15 (1964), 154–69; 16 (1965), 10–24, 71–86, 101–18, 153–77. I would like to thank Judith Entrena, librarian of the Fundación Uriach, for kindly providing me with this article.

57. *Semanario* VIII (1800), 81: "De los efectos de las fricciones con éter acético en los reumatismos, ciática y aun en la gota."

58. *Semanario* X (1801), 201: "Método que se emplea en Lieja para fabricar la sal amoniaco."

59. *Semanario* X (1801), 204: "Por si no se entiende bien la breve descripción que hace Chevremont añadiré que, según yo lo comprendo, se ha de poner en vasos de barro de la hechura de un huevo partes iguales de la sal y el hollín que se ha sacado de los hornos."

60. Eugenio Ortazán y Brunet [Antonia Gutiérrez Bueno], *Diccionario histórico y biográfico de mujeres célebres* (Madrid: Imp. de Cruz González, 1835); Eugenio Ortazán y Brunet [Antonia Gutiérrez Bueno], *Recopilación de lo más interesante que se ha publicado en abril de 1832 en la Gaceta de Francia concerniente al Cólera-Morbo* (Madrid: Pedro Ximénez de Haro, 1832).

61. *Semanario* VII (1800), 402: "Modo de sacar los pollos artificialmente y criarlos cuando hace mucho frío." Her maiden name was Marie-Armande-Jeanne Gacon (1753–1835).

62. Spary, *Feeding France*, 46–48; June K. Burton, *Napoleon and the Woman Question: Discourse of the Other Sex in French Education, Medicine and the Medical Law 1779–1815* (Lubbock: Texas Tech University Press, 2007), 74–78.

63. For example, the priest of Linares disagreed with Chancey, Melón disagreed with Arthur Young in regard to the use of sand in barns floors, and the editors did not agree that an excess of alfalfa could make sheep sick, as was believed by Gilbert from the Paris Academy of Agriculture; *Semanario* VI (1799), 330, "El cura de Linares al Semanario de Agricultura"; *Semanario* X (1801), 21n 2; *Semanario* 6 (1799), 146n 2. See also Serrano, "Making Oeconomic People," 12.

64. *Semanario* XVII (1805), 406: "Al Semanario se debe la publicación de unos buenos elementos de química, botánica, de agricultura y de parte de historia natural, escritos en estilo fácil y acomodados a la comprensión de las mujeres, que enteradas de ellos inspiren a las nuevas generaciones los importantes conocimientos que hasta ahora han estado por desgracia muy apartados de nuestra educación."

65. Chapter 1 in this book; Mary Terrall, "Metaphysics, Mathematics, and the Gendering of Science in Eighteenth-Century France," in *The Sciences in Enlightened Europe*, ed. William Clark, Jan Golinski, and Simon Schaffer (Chicago: Chicago University Press, 1999), 246–71; idem., "Natural Philosophy for Fashionable Readers," in *Books and the Sciences in History*, ed. Marina Frasca-Spada and Nick Jardine (Cambridge, UK: Cambridge University Press, 2000), 239–53. See also: Anthony J. La Vopa, *The Labor of the Mind: Intellect and Gender in Enlightenment Cultures* (Philadelphia: University of Pennsylvania Press, 2017).

66. "Compendio de la chimica acomodado a la instrucción de las mujeres," *Semanario* X (1801), 272; 274–88; 293–303; 304–20; 326–36; 343–49; *Semanario* XI (1802), 67–80; 84–94; 99–112; 115–28; 131–43; 147–56; *Semanario* XII (1802), 50–64; 68–80; 85–94.

67. "Principios de botánica en cartas a una señora": *Semanario* XII (1802), 209–24; 226–39; 242–56; 261–72; 279–87.

68. This was not in the letter format; the latter begins in *Semanario* XIV (1803).

69. *Semanario* XVI (1804), 347: "y no me pidas más, que ya mi débil máquina se haya muy fatigada de trabajo." The treatise is called "Elementos de Historia Natural en Cartas a una Señora" and begins in *Semanario* XVI (1804): 346–51; 362–68; 371–72; *Semanario* XVII (1805), 152–60; 161–75; 178–89; 199–208 (they were not completed).

70. Giuseppe Compagnoni, *La chimica per le donne* (Venezia: Dalla tipografia Pepoliana, Presso Antonio Curti q. Giacomo, 1796); Antonio José Cavanilles, *Descripción de las plantas que D. Antonio Josef Cavanilles demostró en las lecciones públicas del año 1801: precedida de los principios elementales de la botánica* (Madrid: Publishing house unknown, 1802).

71. *Semanario* X (1801), 272, letter 1: "Me dices, amable C., que quieres aprender la química por ser ciencia de moda, y porque siguiendo la moda en el estudio, no se tendrá por el octavo de los vicios capitales."

72. Agustí Nieto-Galán, "The French Nomenclature in Spain: Critical Points, Rhetorical Arguments and Practical Uses," in *Lavoisier in European Context*, ed. Bernadette Bensaude-Vincent and Ferdinando Abbri (Canton MA: Science History, 1995), 173–91; Antonio García Belmar and José R. Bertomeu Sánchez, "Spanish Chemistry Textbooks during the Late Eighteenth Century: Building Up a New Genre of Scientific Literature," in *Universities and Science in the Early Modern Period*, ed. Mordechai Feingold and Victor Navarro (Dordrecht: Springer, 2006), 241–57; Antonio García Belmar and José R. Bertomeu Sánchez, "Pedro Gutiérrez Bueno (1745–1822), los libros de texto y los nuevos públicos de la química en el último tercio del siglo XVIII," *Dynamis: Acta Hispanica ad Medicinae Scientiarumque Historiam Illustrandam* 21 (2001): 351–74; Ramón Gago, "The New Chemistry in Spain," *Osiris* 4 (1988): 169–92. See also the volume edited by Robert Fox and Agustí Nieto-Galán, *Natural Dyestuffs and Industrial Culture in Europe* (Canton, MA: Science History, 1999).

73. *Anales de Historia Natural* (1799) changed its name to *Anales de ciencias naturales* (1801–1804).

74. On the role of spectacles in shaping eighteenth-century science, see the collection of essays in Bernadette Bensaude-Vincent and Christine Blondel, ed., *Science and Spectacle in the European Enlightenment* (Aldershot: Ashgate, 2008).

75. *Diario de Madrid*, February 23, 1799; March 30, 1797, 363.

76. *Diario de Madrid*, October 7, 1797: "Don Francisco Bienvenu, profesor de física, para contentar al público que desea saber si las señoras pueden suscribir y asistir al curso que principiará el lunes 9 del corriente tiene el honor de advertir que entre los suscriptores se cuentan ya varias señoras de esta Corte, como las ha habido de continuo en París y en las varias cortes y ciudades de la Europa en las que dicho don Francisco Bienvenu ha hecho sus experiencias." See John E. Varey, *Cartelera de los títeres y otras diversiones populares de Madrid: 1758–1840; Estudio y documentos* (Madrid: Támesis, 1995), 162–41.

77. Patrice Bret, "Un bateleur de la science: Le 'machiniste-physicien' François Bienvenu et la diffusion de Feingold et Lavoisier," *Annales Historiques de la Révolution Française* 338 (2004): 95–127.

78. For an examination of the private courses of Pedro Gutiérrez Bueno, see "Relación de los ejercicios literarios y méritos y servicios de D. Pedro Gutiérrez Bueno," box 490, exp. 26, Archivo Palacio Real. On Herrgen, see Dolores Parra and Francisco Pelayo, "Christian Herrgen y la institucionalización de la mineralogía en Madrid," *Asclepio* 48 (1996): 163–81, here 179.

79. John Perkins, "Chemistry Courses, the Parisian Chemical World and the Chemical Revolution, 1770–1790," *Ambix* 57, no. 1 (March 2010): 27–47.

80. Giuseppe Compagnoni, *Cartas físico-químicas escritas en italiano por el señor Compagnoni, y traducidas al castellano por Don Josef Antonio Sabater y Anglada*, trans. Antonio Sabater (Barcelona: Oficina de Pablo Nadal, 1802).

81. The letters were addressed to the Countess Marianna Rossi-Gnudi from Ferrara.

82. Marcello Savini, *Un abate libertino, Le memoire autobiografiche e altri scritti di Giussepe Compagnoni* (Lugo: Banca di Romagna, 1988), 234–35.

83. Andrea Cristiani, "Dall'entusiasmo al plagio: Ossidazioni e riduzioni verbali nella chimica divulgativa di Giuseppe Compagnoni," in *Giuseppe Compagnoni: Un intellettuale tra giacobinismo e restaurazione*, ed. Sant Medri (Bologna: Edizioni Analisi, 1993); Stefania

de Toma, "Giuseppe Compagnoni e la divulgazione scientifica nella Chimica per le donne," in *La letteratura italiana a Congresso: Bilanci e prospettive del decennale (1996–2006), Atti del X Congresso dell'ADI*, ed. R. Cavalluzzi et al. (Lecce: Pensa multimedia, 2008), 619–28; Corinna Guerra, "Chimica per la donne," in *Scorci di storia della scienza*, ed. Frank Martin L. and C. Pogliano (Pisa: Pisa University Press, 2010), 127–39.

84. Vicenzo Dandolo, *Fondamenti della scienza chimico-fisica applicati alla formazione de corpi ed al fenomeno della natura* (Milan: Tipografia Milanese di Tosi e Nobile, 1802); Marco Beretta, "Italian Translations of the *Méthode de Nomenclature Chimique* and the *Traité Élémentaire de Chimie*: the Case of Vicenzo Dandolo," in Bensaude-Vincent and Abbri, *Lavoisier in European context*, 224–37, here 232.

85. *Fondamenti*, 29: "quadro filosofico-chimico dell'universo"; *Fondamenti*, "Discorso Preliminare," i–xxv, xii.

86. Giuseppe P. Poli, *Elementi di fisica sperimentali* . . . (Venice: Publishing house unknown, 1796).

87. According to Cristiani, this occurred from the letter XX. Cristiani, "Dall' entusiasmo al plagi," 163.

88. Cristiani, "Dall'entusiasmo al plagio," 161: "Semplice e facilissimo, e con precisione dottrinale." There were three Italian editions: 1796, 1797, and 1805.

89. Compagnoni, *Cartas*, 1.

90. On the uses of gallantry in philosophical discussions, see: La Vopa, *Labor of the Mind*.

91. Compagnoni, *Cartas*, 19: "y Lavoisier ha sido en la Química lo que doscientos años atrás fue Galileo en la Física . . . y destruyendo de un golpe todas las antiguas fórmulas, por la mayor parte falsas, inexactas, absurdas o imperfectas, inventó un lenguaje correspondiente a los nuevos principios, del cual me dice V. M."

92. Beretta, "Italian translations of the *Méthode*."

93. The first eight letters were written between 1771 and 1773 and circulated widely in manuscripts, and then in posthumous editions. In 1785 they were translated into English, along with twenty-four additional letters by Thomas Martyn, a botanical professor at Cambridge. This edition was again translated into French in 1800 and 1802 and magnificently illustrated by Pierre-Joseph Redouté in 1805. Cited here as *Jean-Jacques Rousseau's Letters on the Elements of Botany: Addressed to a Lady*, trans. Thomas Martyn (London: White, 1794), 19; see also Jean-Jacques Rousseau, "Essais élémentaires sur la botanique [Introduction; lettres élémentaires]," in Jean-Jacques Rousseau, *Collection complète des oeuvres de J.-J. Rousseau, . . . publiées par Du Peyrou* (London: Publishing house unknown, 1780–1782). For scholarship on the letters, see Roy McMullen, "Introduction," in *Botany: A Study of Pure Curiosity; Botanical Letters and Notes towards a Dictionary of Botanical Terms by Jean-Jacques Rousseau* (London: Felix Gluck Press, 1979). See also Ann B. Shteir, *Cultivating Women, Cultivating Science: Flora's Daughters and Botany in England, 1760 to 1860* (Baltimore: Johns Hopkins University Press, 1996); Palmira Fontes da Costa, "Women and the Popularisation of Botany in Early Nineteenth-Century Portugal: The Marquesa de Alorna's Botanical Recreations," in Papanelopoulou, Nieto-Galán, and Perdiguero, *Popularizing Science and Technology*, 43–63.

94. See, for instance, Lisbet Koerner, *Linnaeus: Nature and Nation* (Cambridge, MA: Harvard University Press, 1999); Schiebinger and Swan, *Colonial Botany*; Yota Batsaki, Sarah Burke Cahalan, and Anatole Tchikine, eds., *The Botany of Empire in the Long Eighteenth Century* (Washington, DC: Dumbarton Oaks Research Library and Collection, 2016); Daniela Bleichmar, *Visible Empire: Botanical Expeditions and Visual Culture in the Hispanic Enlightenment* (Chicago: University of Chicago Press, 2012).

95. Cavanilles, *Descripción de las plantas.*

96. Priscilla Wakefield, *An Introduction to Botany: In a Series of Familiar Letters, with Illustrative Engravings* (Dublin: Thomas Burnside, 1796). There is a useful introduction to Priscilla Wakefied in Ann B. Shteir, "Priscilla Wakefield's Natural History Books," in *From Linnaeus to Darwin: Commentaries on the History of Biology and Geology,* ed. Alwayne Wheeler and James H. Price (London: Society for the History of Natural History, 1985), 29–50.

97. Staffan Müller-Wille, "Linnaeus and the Love Lives of Plants," in *Reproduction: From Antiquity to the Present Day,* ed. N. Hopwood, R. Flemming, and L. Kassell (Cambridge, UK: Cambridge University Press, 2018), 305–18; Erasmus Darwin, *The Botanic Garden: A Poem in Two Parts; Part I Containing the Economy of Vegetation, Part II The Loves of the Plants* (London: 1789–1791).

98. The translation was not published at the time, but it might have circulated as a manuscript.

99. *Semanario* XII (1802), 222: "Te ruego estimada C. que no pierdas la esperanza de divertirte en la anatomía y descripción de las plantas"; *Semanario* XII (1802), 237: "Me admira o virtuosa C. la constancia con que examinas las flores"; *Semanario* XII (1802), 245: "Tienes razón en decir que son muy secas mis cartas cuando sólo contienen listas de nombres"; *Semanario* XII (1802), 256: " Ya voy a dar fin a esto elementos escabrosos y áridos . . . no te cuento a ti entre estos indolentes pues sueles repetir "por estas asperezas se camina de la inmortalidad a la alta cumbre, do nunca arriba quin de aquí declina."

100. *Semanario* XII (1802), 215: "Ya veo que te repugnará toda esta cáfila de nombres que te pongo abajo para que te sean mis cartas menos enfadosas, y para que solo recorras estas listas cuando vayas a dar los nombres a las partes del vegetal que estés reconociendo."

101. For instance, the one by Casimiro Gómez Ortega and Antonio Palau y Verdera, *Curso elemental de botánica teórico y práctica, dispuesto para la enseñanza del Real Jardín Botánico de Madrid* (Madrid: Imprenta Real, 1785).

102. *Semanario* XII (1802), 256. Linnaeus classified, roughly, the plants into twenty-four classes in accordance with the number of stems that the flower had. Thus, if the flower had only one stem, it was a Monandria; if it had two, it was a Diandra, etc.; this delineated the class. The second name of the plant designates the order and depends on the number of pistils (Monogynia, Digynia, Tryginia, etc.). The *Semanario* presented the Linnaeus system in a form modified by the botanical garden.

103. Antonio Palau y Verdera, *Parte práctica de Botánica del caballero Cárlos Linneo que comprehende las clases, órdenes, géneros, especies y variedades de las plantas, con sus Caracteres genéricos y específicos* (Madrid: Imprenta Real, 1784–1788).

104. *Semanario* XII (1802), 267: "Sal con tu familia al campo por todo el día a ver cómo buscas las plantas y las vas poniendo en sus clases, órdenes, especies . . . supongo que llevarás el lente y demás instrumentos y unos cuadernillos de papel para meter entre sus hojas varias plantas con sus flores."

105. The excessive praise of Cavanilles throughout the text as well as the light-hearted style suggest that the author of the adaptation was Melón, and not Cavanilles as was stated in the authors' index.

106. *Semanario* XII (1802), 226: "acuérdate de la envidia que te dio nuestro insigne pintor Goya cuando le viste embelesado con las pinturas de la sacristía del Escorial"; *Semanario* XII (1802), 279: "He tenido mucha complacencia en que dieses a conocer tus progresos en la botánica en el bellísimo jardín de la Salm, aquella señora a quien tanto respetas por su talento, buen gusto y circunspección."

107. *Semanario* XII (1802), 227. See also Domingo A. Madulid, "The Life and Work of Luis Née, Botanist of the Malaspina Expedition," *Archives of Natural History* 16 (1989): 33–48.

108. *Semanario* XIII (1802), 248: "Si tú asistieras a las lecciones del sabio botánico, cuyos elementos quieres que te vaya dando en mis cartas . . . en otros países concurren a estas aulas, y a las de Física y Química, muchas personas de tu sexo: entre nosotros estáis como desterradas de estas concurrencias . . . El no poder tu concurrir a la enseñanza en el jardín por no singularizarte siendo tu la primera."

109. *Semanario* XII (1802), 286: "Aquí tienes reducido á un breve tratado lo que deseabas saber sobre la botánica para instruir a tu hija. Si hubiese muchos que se aplicasen con la constancia que tú á adquirir estos conocimientos, es indecible lo que pudiera adelantar la agricultura, la industria, la medicina, la historia natural, y los medios de subsistir y de enriquecernos; pero si no sabemos conocer las preciosas producciones que tal vez pisamos; si no distinguimos bien ni aun las especies de trigo que cultivamos para explicarlas á otro; si desentendiéndonos de todo lo que nos rodea pasamos la vida pensando, meditando y estudiando en lo que no tiene relación con nuestra existencia y bienestar, viviremos ignorantes, pobres, y despreciados de las naciones que han sabido sacar de las ciencias naturales la riqueza y el poder con que amenazan subyugar a las demás."

110. Luisa Gómez Carabaño, *Del cultivo de las flores que provienen de cebolla* (Madrid: Tomás Alban, 1825).

111. Antonio Ponz, *Viaje de España o Cartas en que se da noticia de las cosas mas apreciables y dignas de saberse, que hay en ella*, 18 vols. (Madrid: Imprenta Vda. de Ibarra, 1772–1794), 13:773: "y acaso tendrá Dios dispuesto que las grandes Señoras efectúen en beneficio de su patria lo que no ha sido concedido a tantos hombres. Todo será que lo reflexionen y paren la consideración de cuan buena moda seria esta sobre cuantas pueden inventarse, y digna de registrarse en los Anales de la Nación para gloriosa memoria de las mismas."

112. Ponz, *Viaje*, 13:lxx.

113. Josefa Alonso de Pimentel (1752–1834), twelfth Countess-Duchess of Benavente, better known today by the title acquired through her marriage, ninth Duchess of Osuna. See chapter 1.

114. Ponz, *Viaje*, 13:lxxj: "La huerta y plantación que esta última Señora [Duchess Osuna] ha hecho hacer en el Lugar de la Alameda, y extensión que piensa darle para hacer más deleitable, útil y frondoso aquel terreno, podrá ser en pocos años uno de los sitios mas deleitables que puedan verse."

115. Carmen Añón Feliu, *"El Capricho" de la Alameda de Osunas* (Madrid: Fundación Caja Madrid, 2001).

116. Juan Antonio Hernández de Larrea, "Carta a Doña Josefa Amar diciendo su parecer sobre el discurso antecedente," *Memorial literario* 32 (August 1786), 430–38: "La Excma. Señora Marquesa de Peñafiel . . . hace fabricar allí una Granja suntuosa, que será Seminario del gusto acerca de los conocimientos y ensayos agrónomos y una escuela que instruye en la ciencia del campo más digna y apreciable que otras instituidas solamente para gritar y disputar sin ventaja alguna de la República."

117. Añón Feliu, *El Capricho*, 49–50.

118. Map 5, Consejos, Archivo Historico Nacional. Nowadays, it counts 14 ha, but was bigger at the time. On the Boutelau family, see Añón Feliu, *El Capricho*, 64. Pablo's sons were the future botanists Claudio and Esteban.

119. Jose Luis Sancho, "Aranjuez y el arte del jardín durante el reinado de Carlos III," *Reales Sitios* 98 (1988): 49–59.

120. In 1787 the duchess contracted the French gardener Jean Baptiste Mulot, who had worked in the Petit Trianon and is known today for imitating a picturesque farming landscape.

121. See Williamson, *Polite Landscapes*.

122. See Richardson, *Arcadian Friends*.

123. For a discussion on social imaginaries and their relation to practices and ideal futures, see Roberts, "Practicing Oeconomy," 135–37. She argues for the "the future-oriented character of such imaginaries, which include beliefs and expectations about how the present ought to stretch out into the future" and relates them to the actual practices of social groups, such as the improvers of economic societies.

124. William Chamber, *Designs of Chinese Buildings, Furniture, Dresses, Machines, and Utensils; Engraved by the Best Hands, from the Originals Drawn in China . . . To Which Is Annexed, a Description of their Temples, Houses, Gardens, &c.* (London: Publishing house unknown, 1757), 15–19.

125. Juan F. Remón Fernández, "The Alameda of the Duchess of Osuna: A Garden of Ideas," *Journal of Garden History* 13, no. 4 (1993): 224–40; Pedro Navascués, "La Alameda de Osuna: Una villa suburbana," *Pro-Arte* 2 (1975): 7–26.

126. See chapter 1.

Conclusion

1. Lisbet Koerner, "Women and Utility in Enlightenment Science," in "Gender and Early Modern Science Cluster," ed. Mary Terrall, special issue, *Configurations* 3, no. 2 (Spring 1995): 223–55, here 235.

2. For an updated discussion of the historiography, see Sebastian Felten and Christine von Oertzen, "Bureaucracy as Knowledge," *Journal for the History of Knowledge* 1 (2020): 1–8.

3. Jonathan R. Topham, "The Scientific, the Literary and the Popular: Commerce and the Reimagining of the Scientific Journal in Britain, 1813–1825," *Notes and Records of the Royal Society of London* 70, no. 4 (December 2016): 305–24.

4. Sally Shuttleworth, "Constructing Scientific Communities: Citizen Science," *Lancet* 385, no. 9987 (2015): 2568.

BIBLIOGRAPHY

Manuscripts are cited fully in the notes and are not included in this list. I cite the edition that I have consulted, which is not always the first edition.

Archives

Archivo Histórico Nacional (AHN)
Archivo Real Academia de Medicina (RAM)
Archivo Regional de la Comunidad de Madrid (ARM)
Archivo Sociedad Ecónomica Matritense (SEM)
Biblioteca Nacional de España (BN)

Primary Sources

An Account of the Hospital for the Maintenance and Education of Exposed and Deserted Young Children. London, 1749.

Amar y Borbón, Josefa. *In Defence of Women.* Translated by Joanna M. Barker. Cambridge, UK: Salisbury House, 2018.

Amar y Borbón, Josefa. *Discurso sobre la educación física y moral de las mujeres.* Madrid: Benito Cano, 1790.

Anacreonte. *Obras de Anacreónte.* Translated by Joseph and Bernabé Canga Argüelles. Madrid: Imprenta de Sancha, 1795.

Aparato, ceremonias, y solemnidades que se executaron en la entrada que hizo en la Real Universidad de Alcalá la Excelentísima Señora Doña María Isidra de Guzmán. N.p.: Publishing house unknown [1785?].

Armstrong, George. *An Account of the Diseases Most Incident to Children.* London: T. Cadell, 1771.

Bary, René. *La fine philosophie, accomodee à l'intelligence des dames.* Paris: Chez Simeon Piget et chez l'autheur, 1660.

Berquin A., *Introduction familière à la connoissance de la nature, traduction libre de l'anglois de Mrs. Trimmer.* Paris: Bureau de l'Ami des enfans, 1784.

Bilbao, Antonio. *Destrucción y conservación de los expósitos: modo breve de poblar la España.* Málaga: Don Félix de Casas y Martínez, 1790.

Bonells, Jaume. *Perjuicios que acarrean al genero humano y al Estado las madres que rehusan criar á sus hijos, y medios para contener el abuso de ponerlos en ama.* Madrid: Miguel Escribano, 1786.

Campe, Joachim Heinrich. *El Nuevo Robinson, historia moral reducida a diálogos para la instrucción y entretenimiento de niños y jóvenes de ambos sexos.* Translated into Spanish by Tomás de Iriarte. Madrid: Benito Cano, 1789.

Campomanes, Pedro Rodríguez de. *Discurso sobre el fomento de la industria popular.* Madrid: Imprenta de Sancha, 1774.

Canga-Argüelles, Joseph. *Diccionario de Hacienda*. London: M. Calero, 1826–1827.

Canga Argüelles, Joseph. *Elementos de la ciencia de la Hacienda*. London: A. Macintosh, 1825.

Carolo Tertio regi, catholico, semper, augusto pio . . . obtinenda in philosophia, et humanioribus litteris doctorali laurea in maximo universitatis complutensis theatre. Madrid: Joachim Ibarra, [1785?].

Castro-Terreño, Countess of. *Elogio de la reyna nuestra señora*. Madrid: Imprenta Real, [1801?].

Catálogo de los libros impresos por D. Antonio de Sancha, [Madrid?]: publishing house unknown, [1772 and 1800].

Cavanilles, Antonio José. *Descripción de las plantas que D. Antonio Josef Cavanilles demostró en las lecciones públicas del año 1801: Precedida de los principios elementales de la botánica*. Madrid: publishing house unknown, 1802.

Cavazzoni Zanotti, Francesco Maria. *Compendio de la Filosofia Moral, traducido del Italiano por la Marquesa de Espeja*. Madrid: Joaquín Ibarra, 1785.

Caveda, Rita. *Cartas selectas de una señora a una sobrina suya, entresacadas de una obra inglesa*. Madrid: Garcia y Compañía, 1800.

Chamber, William. *Designs of Chinese Buildings, Furniture, Dresses, Machines, and Utensils; Engraved by the Best Hands, from the Originals Drawn in China . . . To Which Is Annexed, a Description of Their Temples, Houses, Gardens, &c.* London: publishing house unknown, 1757.

Chaptal, Jean-Antoine. *De l'industrie française*. Paris: Corréard, 1819.

The Children's Magazine; or, Monthly Repository of Instruction and Delight. London: John Marshall, 1799–1800.

Clavijo y Fajardo, José. *Prólogo a la traducción de la "Historia natural" del Conde de Buffon. Estudio preliminar de José Luis Prieto Pérez*. La Orotava: Fundación Canaria Orotava de Historia de la Ciencia, 2001.

Clemencín, Diego. *Discurso leído en la abertura del examen público de las señoras doña Josefa y doña Joaquina Girón y Pimentel, y de los señores d. Francisco y d. Pedro, sus hermanos, hijos de los excelentísimos señores Duques de Osuna, Condes-Duques de Benavente por don Diego Clemencín, el día 20 de abril de 1797*. Madrid: Benito Cano, [1797?].

Clemencín, Diego. *Lecciones de gramática y ortografía castellana. Edición facsimilar de la edición de Madrid*. Madrid: Imprenta de D. Miguel de Burgos, 1842.

Clemencín, Diego. *Proyecto para la educación del Exmo. Sr. Marqués de Peñafiel, y del Sr. Príncipe de Anglona. Dirigido a sus padres, los Exmos. Duques de Osuna, Condes-Duques de Benavente.*

Compagnoni, Giuseppe. *Cartas fisico-químicas escritas en italiano por el señor Compagnoni, y traducidas al castellano por Don Josef Antonio Sabater y Anglada*. Translated into Spanish by José Antonio Sabater. Barcelona: Oficina de Pablo Nadal, 1802.

Compagnoni, Giuseppe. *La chimica per le donne*. Venice: Tipografia Pepoliana, Presso Antonio Curti q. Giacomo, 1796.

Constituciones, ordenanzas y reglamentos del Real Hospicio . . . de la Ciudad de Badajoz. Madrid: Josef Collado, 1804.

Copia y recolección de los papeles que, en prosa y verso, han dirigido algunos doctos ingenios de esta ciudad, en debido aplauso del desempeño que en sus actos literarios de los días 19, 22 y 24 de el mes próximo pasado, ejecutó la señora Doña María del Rosario Cepeda. Cádiz: Imprenta Real de Marina, 1768.

Dandolo, Vizenzo. *Fondamenti della scienza chimico-fisica applicatiu alla formazione de corpi ed al fenomeni della natura*. Milan: Tipografia Milanese di Tosi e Nobile, 1802.

Darwin, Erasmus. *The Botanic Garden: A Poem in Two Parts; Part I Containing the Economy of Vegetation, Part II The Loves of the Plants*. London: publishing house unknown, 1789–1791.

Delgado, Antonio. *Por el magisterio en artes liberales que se confirió en la Real Universidad de Alcalá en 6 de junio de 1785 a la ilustrísima señora doña María Isidra de Guzmán y de la Cerda. . . .* Madrid: Antonio Delgado, 1785.

Diario Curioso, Erudito, Económico y Comercial. Madrid: 1786–1787.

Diderot, Denis. "Rustic Economy." In *The Encyclopedia of Diderot & d'Alembert Collaborative Translation Project*. Translated by Ann-Marie Thornton. Ann Arbor: Michigan Publishing, University of Michigan Library, 2013.

Díez de la Cortina y Morales, Josefa. *Elogio de la Reyna Nuestra Señora . . . leído en la junta publica de distribución de premios en 18 de diciembre de 1799*. Madrid: Imprenta Real, 1800.

Dubroca, Jean François. *Conversaciones de un padre con sus hijos sobre la historia natural*. Translated into Spanish by Manuel María de Ascargorta. Madrid: Imprenta Real, 1802–1803.

Ducray-Duminil, François-Guillaume, *Ambrose and Eleanor or the Adventures of Two Children Deserted on an Uninhabited Island*. London: R. & L. Peacock, 1796.

Ducray-Duminil, François-Guillaume. *Lolotte et Fanfan, ou les Aventures de deux enfans abandonnés dans une isle déserte, rédigées . . . sur des manuscrits anglais*, 2 vols. Paris: Maradan, 1788.

Ducray-Duminil, François-Guillaume, *Los dos Robinsones o aventuras de Carlos y Fanny*. Madrid: Justo de la Barra, 1797.

Espinosa, Antonio. *El ceremonial de estrados y critica de visitas: Obra útil, curiosa, y divertida*. Madrid: Antonio Espinosa, 1789.

Espíritu de los mejores diarios literarios que se publican en Europa. Madrid: Manuel González y Antonio Espinosa, 1787–1790.

Estatutos de la Junta de Socias de Honor y Mérito de la Real Sociedad Económica de Madrid. Madrid: Sancha, 1794.

Estienne, Robert. "Eloge historique de monsieur l'abbé Pluche." In Antoine-Noel Pluche, *Concorde de la géographie des différens âges*. Paris: Chez les frères Estienne, 1764.

Examen al que se presentará Doña Pascuala Caro y Sureda. Valencia: Benito Monfort, 1781.

Examen público de las señoras Doña Josefa y Doña Joaquina Girón y Pimentel, y de los señores Francisco y D. Pedro sus hermanos hijos de los excelentísimos señores Duques de Osuna Condes-Duques de Benavente, el día 20 de abril de 1797. Madrid: Benito Cano, 1797.

Extracto de las Juntas Generales celebradas por la Real Sociedad Bascongada de los Amigos del Pais, en la Villa de Vergara, por Octubre de 1789. Vitoria: Baltasar de Manteli, 1789.

Fénelon, François de Salignac de La Mothe. *Instructions for the Education of a Daughter*. London: Printed by and for J. Ferraby, 1797.

Fénelon, François de Salignac de La Mothe. *Tratado de la educación de las hijas*. Madrid: Viuda de Eliseo Sánchez, 1769.

Ferris, Samuel. *A Dissertation on Milk*. London: John Abraham, 1785.

Fillassier, Jean-Jacques. *Eraste ou l'ami de la jeunesse*. Paris: Vincent, 1773.

Fontenelle, Bernard Bovier de. *A Discourse of the Plurality of Worlds*. Dublin: Andr. Crook and Sam. Helsham, 1687.

Fontenelle, Bernard Bovier de. *Conversaciones sobre la pluralidad de los mundos*. Madrid: Imp. de Villalpando, 1796.

Fontenelle, Bernard Bovier de. *Conversations on the Plurality of Worlds*. London: Printed for A. Bettesworth and E. Curll, 1715.

Fontenelle, Bernard Bovier de. *Entretiens sur la pluralité des mondes*. Paris: Veuve C. Blageart, 1686.

Foronda, Valentín de. *Lecciones ligeras de química*. Madrid: Imprenta de Gonzalez, 1791.

François Fourcroy, Antoine. *Diario de los nuevos descubrimientos de todas las ciencias físicas, que tienen alguna relacion con las diferentes partes del arte de curar*. 2 vols. Madrid, Imprenta de Sancha, 1792–1793.

García, Santiago. *Instituciones sobre la crianza física de los niños expósitos*. Madrid: Imprenta de Vega y Compañía, 1805.

Gazeta de los niños. Madrid: Imprenta de Sancha, 1798–1800.

Gazeta de Madrid. 1697–1936.

Genlis, Stéphanie Félicité, Comtesse de. *Las veladas de la quinta o Novelas e historias sumamente útiles para que las madres de familia, a quienes las dedica la autora, puedan instruir á sus hijos, juntando la doctrina con el recreo*. Madrid: Imprenta M. González, 1788.

Genlis, Stéphanie Félicité, Comtesse de. *Les veillées du château*. Paris: M. Lambert et F.-J. Baudouin, 1782.

Gómez Carabaño, Luisa. *Del cultivo de las flores que provienen de cebolla*. Madrid: Tomás Alban, 1825.

Gómez Ortega, Casimiro, and Antonio Palau y Verdera. *Curso elemental de botánica teórico y práctica, dispuesto para la enseñanza del Real Jardín Botánico de Madrid*. Madrid: Imprenta Real, 1785.

González, Teresa. *El estado del cielo para el año de 1778, arreglado al meridiano de Madrid*. Madrid: Manuel Martín, [1778?].

González Cañaveras, Juan Antonio. *Plan de educacion o Exposición de un nuevo método para estudiar las lenguas, geografía, cronología, historia*. Cadiz: Viuda de D. Antonio de Alcántara, 1784.

Griselini, Francesco. *Discurso sobre el problema de si corresponde a los párrocos y curas de las aldeas el instruir a los labradores en los buenos elementos de la economía campestre*. Zaragoza: Blas Miedes, 1784.

Guzmán y La Cerda, María Isidra Quintina de. *Oración del género eucarístico . . . a la Real Sociedad de Amigos del País*. Madrid: Antonio de Sancha, 1786.

Guzmán y La Cerda, María Isidra Quintina de. *Oración del género eucarístico, que hizo a la Real Academia Española . . . por Socia de dicha Real Academia*. Madrid: Joaquín Ibarra, 1785.

Iriarte, Tomás. *Lecciones instructivas sobre la historia y la geografia: Obra póstuma*. 3 vols. Madrid: Imprenta Real, 1794.

Jauffret, Louis-François. *Éléments de zoographie, ou l'histoire des animaux considérés relativement au degré d'étendue des régions que chaque espèce occupe sur la surface du globe*. Paris: Demoraine, n.d.

Jauffret, Louis-François. *Jeux zoologiques et géographiques*. Paris: Leclère, 1799.

Jauffret, Louis-François. *Las gracias de la niñez y placeres del amor maternal*. Madrid: Imprenta Real, 1804.

Jauffret, Louis-François. *Les charmes de l'enfance et les plaisirs de l'amour maternel*. Paris: Moutard, 1791.

Jauffret, Louis-François. *Les merveilles du corps humain*. Paris: A. J. Dugour et Durand, 1798.

Jauffret, Louis-François. *Les voyages de Rolando et de ses compagnons de fortune autour du monde*. Leipzig: J. C. D. Sinner, 1803.

Jauffret, Louis-François. *Los viajes de Rolando y de sus compañeros de fortuna alrededor del mundo*. Madrid: Imprenta Real, 1804.

Jauffret, Louis-François. *Voyage au jardin des Plantes, contenant la description des galeries d'histoire naturelle, des serres . . . de . . . l'école de botanique, avec l'histoire des . . . animaux de la ménagerie nationale.* Paris: C. Houel, 1797.

Llampillas, Francisco Xavier. *Ensayo histórico-apologético de la literatura española contra las opiniones preocupadas de algunos escritores modernos* italianos. 7 vols. Translated into Spanish by Josefa Amar y Borbón. Zaragoza: Blas Miedes, 1782–1786.

Llampillas, Francisco Xavier. *Respuesta del señor abate Lampillas a los cargos recopilados por el señor abate Tiraboschi en su carta al señor abate N.N. sobre el ensayo histórico apologético de la literatura Española, traducid traducido del italiano por D. Josefa Amar y Borbón*. Zaragoza, Blas de Miedes, 1786.

LeMaur, Carlos. *Discurso sobre la astronomía ó introduccion al conocimiento de los fenómenos astronómicos, sus leyes, su causa y su aplicacion à los usos de la vida civil.* Madrid: Francisco Xavier Garcia, 1762.

Letourneux, Nicolas. *Instrucciones cristianas sobre el sacramento del matrimonio.* Barcelona: Bernardo Pla, 1774.

Luque, Luis. *Arte de partida doble ilustrado.* Cádiz: Calle de la Compañía, 1783.

Martínez de la Mata, Francisco. *Suplemento al apendice de la educacion popular: Contiene dos discursos de Francisco de la Mata . . . de la Orden Tercera de la Penitencia lo publica con algunas notas Don Josef Antonio Canga Argüelles y Cifuentes Prada Martínez de la Mata, Francisco.* Madrid: Imprenta de Sancha, 1794.

Martínez de la Torre, Fausto. *Plano de la Villa y Corte de Madrid: En sesenta y cuatro láminas, que demuestran otros tantos barrios en que está dividida, con los nombres de todas sus plazuelas y calles.* Madrid: Imprenta D. Joseph Doblado, 1800.

Mercurio de España. Madrid: Imprenta Real, 1784–1830.

Memorial Literario Instructivo y Curioso de la Corte de Madrid. Madrid: Imprenta Real, 1784–1791.

Minerva de la Juventud Española. Madrid: publishing house unknown, 1833–1834.

Mora, Pascual. *El hombre en la primera época de su vida.* 3 vols. Madrid: Martínez Dávila, 1827.

Morales, Joseph I. *Comentario de D. Joseph Isidoro Morales al Exc. Señor D. Joseph de Mazarredo sobre la enseñanza de su hija publicado por Don Tadeo Bravo de Rivero.* Madrid: Gabriel de Sancha, 1796.

Murcia, Pedro Joaquín de. *Discurso político sobre la importancia y necesidad de hospicios, casas de expósitos y hospitales que tienen todos los estados y particularmente España.* Madrid: Imprenta Viuda de Ibarra, 1798.

Ortazán y Brunet, Eugenio [Antonia Gutiérrez Bueno]. *Diccionario histórico y biográfico de mujeres célebres.* Madrid: Imp. de Cruz González, 1835.

Ortazan y Brunet, Eugenio [Antonia Gutiérrez Bueno]. *Recopilación de lo más interesante que se ha publicado en abril de 1832 en la Gaceta de Francia concerniente al Cólera-Morbo.* Madrid: Pedro Ximénez de Haro, 1832.

Palau y Verdera, Antonio. *Parte práctica de Botánica del caballero Cárlos Linneo que comprehende las clases, órdenes, géneros, especies y variedades de las plantas, con sus caracteres genéricos y específicos.* Madrid: Imprenta Real, 1784–1788.

Parmentier, Antoine Augustin. *Bibliothèque universelle des dames: Économie rurale et domestique.* 8 vols. Paris, Gaspard-Joseph Cuchet, 1788–1797.

Píndaro. *Obras de Píndaro.* Translated into Spanish by Joseph and Bernabé Canga Argüelles. Madrid: Imprenta de Sancha, 1798.

Pluche, Nöel-Antoine. *Le spectacle de la nature, ou Entretiens sur les particularités de l'histoire naturelle qui ont paru les plus propres á rendre les jeunes-gens curieux*, 8 vols. Paris: Vve Estienne, 1732–1750.

Poli, Giuseppe P. *Elementi di fisica sperimentali. . . .* Venice: publishing house unknown, 1796.

Ponz, Antoni. *Viaje de España o Cartas en que se da noticia de las cosas mas apreciables y dignas de saberse, que hay en ella*. 18 vols. Madrid: Imprenta Vda. de Ibarra, 1772–1794.

Pougens, Charles de. *Mémoires et souvenirs de Charles de Pougens, chevalier de plusieurs ordres. . .* . París: Librarie de H. Fournier Jeune, 1834.

Puig, Antonio Francisco. *Elementos generales de policía, escritos por el señor Juan Henrique Gottlobs de Justi*. Barcelona: Pi Ferrer, 1784.

Real Cédula de S. M. por la que se manda observar el Reglamento inserto para la Policía general de Expósitos de todos sus dominios. Available online as "Cédula 1796–12–11" at http://www.liburuklik.euskadi.eus/handle/10771/26927.

Real Orden de S.M. por la que se sirve aprobar los Estatutos de la Asociación de Señoras establecida en esta Ciudad para ejercitar la caridad en el cuidado de los Niños Expósitos. Málaga: Imprenta de Luis de Carreras y Ramón, 1796.

Relación de los ejercicios literarios que la señora doña María del Rosario Cepeda y Mayo, [. . .], *actuó los días 19, 22 y 24 de septiembre del presente año*. Cádiz: Imprenta Manuel Espinosa de los Monteros, 1768.

Rollin, Charles. *De la manière d'enseigner et d'étudier les Belles Lettres: Par rapport à l'esprit & au cœur*. Paris: Chez Jacques Estienne, 1726–1728.

Rollin, Charles. *Modo de enseñar, y estudiar las Bellas Letras*. Madrid: Impr. del Mercurio, por Joseph de Orga, 1755.

Rousseau, Jean-Jacques. "Essais élémentaires sur la botanique [Introduction; lettres élémentaires]." In Jean-Jacques Rousseau, *Collection complète des oeuvres de J.-J. Rousseau, . . . publiées par Du Peyrou*. (London: publishing house unknown, 1780–1782).

Rousseau, Jean-Jacques. *Jean-Jacques Rousseau's Letters on the Elements of Botany: Addressed to a Lady*. Translated by Thomas Martyn. London: White, 1794.

Rowley, William. *A Practical Treatise on Diseases of the Breasts of Women . . . a Plate for a New Invented Machine, for More Commodiously Drawing the Milk from the Breasts*. London: Newbery, 1772.

Rozier, François. *Curso completo ó Diccionario universal de agricultura: Teórica, práctica, económica, y de medicina rural y veterinaria. . . .* 16 vols. Madrid: Imprenta Real, 1797–1803.

Salvà i Campillo, Francesc, and Francesc Santapons. *Disertación sobre la explicación y uso de una nueva máquina para agramar cáñamos y linos inventada por los doc. en medicina*. Madrid: Imprenta Real, 1784.

Saura, Pedro. "Tratado y explicación de la teta artificial y modo de criar los niños sin amas." *Semanario económico* 8, nos. 13 and 20 (1767).

Sancho, Jose Luis. "Aranjuez y el arte del jardín durante el reinado de Carlos III." *Reales Sitios* 98 (1988): 49–59.

Semanario de agricultura y artes dirigido a los párrocos. Madrid: Imprenta de Villalpando, 1797–1808.

Sempere y Guarinos, Juan. *Ensayo de una biblioteca española de los mejores escritores del reinado de Carlos III*. Madrid: Imprenta Real, 1785–1789.

Sempere y Guarinos, Juan. *Reflexiones sobre el buen gusto en las ciencias y en las artes. Traducción libre*. Madrid: publishing house unknown, 1872.

Thomas, Antoine-Léonard. *Essai sur le caractère, les moeurs et l'esprit des femmes dans les différens siècles*. Paris: Moutard, 1772.

Thomas, Antoine-Léonard. *Historia o pintura del talento, carácter y costumbres de las mujeres en los diferentes siglos*. Madrid: Miguel Escribano, 1773.

Tom Telescope: The Philosophy of Tops and Balls or the Newtonian System of Philosophy. London: J. Newbery, 1761.

Toruneaux, Nicolás. *Instrucciones cristianas sobre el sacramento del matrimonio y sobre las ceremonias con que la Iglesia le administra*. Translated into Spanish by María de Sales Portocarreño, Countess of Montijo. Barcelona: Bernardo Pla, 1774.

Trespalacios y Mier, Juan Antonio. *Discurso sobre que los niños expósitos consigan en las inclusas el fin de estos establecimientos*. Madrid: Imprenta de Villalpando, 1798.

Valcárcel, José Antonio. *Agricultura general y gobierno de la casa de campo*. 10 vols. Valencia: Estevan Dolz, 1765–1795.

Valdés, Pedro Díaz de. *El padre de su pueblo o medios para hacer temporalmente felices a los pueblos con el auxilio de los señores curas párrocos*. Vitoria: publishing house unknown, 1793.

Valdés, Pedro Díaz de. *Memoria premiada por la Real Sociedad Bascongada e impresa en Victoria en 1793. Reimpresa ahora, con un discurso previo y algunas notas*. Barcelona:Manuel Texero, 1806.

Verdadera y sucinta noticia de las ocupaciones y buenos servicios que hacen a la humanidad las Hijas de la Caridad. Madrid: publishing house unknown, 1813.

Wakefield, Priscilla. *An Introduction to Botany: In a Series of Familiar Letters, with Illustrative Engravings*. Dublin: Thomas Burnside, 1796.

Xenofonte. *La economía y los medios de aumentar las rentas públicas de Atenas*. Translated into Spanish by Ambrosio Ruiz Bamba. Madrid: Benito Cano, 1786.

Young Gentleman's and Lady's Magazine. London: J. Walker, 1799–1800.

Young, Arthur. *Travels in France during the Years 1787, 1788, 1789*. London: W. Richardson, Royal-Exchange, 1794.

Secondary Sources

Acereda, Alberto. *La Marquesa de Fuerte-Híjar: Una dramaturga de la Ilustración*. Cádiz: Universidad de Cádiz, 2000.

Acereda, Alberto. "Una figura relegada de la Ilustración: La Marquesa de Fuerte-Híjar y su Elogio a la Reina (1798)." *Cuadernos de Investigación Filológica* 23–24 (1997–1998): 195–212.

Adams, Christine. *A Taste for Comfort and Status: A Bourgeois Family in Eighteenth-Century France*. University Park: Pennsylvania State University Press, 2000.

Agar, Jon. *The Government Machine: A Revolutionary History of the Computer*. Cambridge, MA: MIT Press, 2003.

Aguilar Piñal, Francisco. *La España del absolutismo ilustrado*. Pozuelo de Alarcón: Espasa, 2005.

Akkerman, Nadine, and Brigit Houben, eds. *The Politics of Female Households: Ladies-In-Waiting across Early Modern Europe*. Leiden: Brill, 2014.

Albarrán, Virginia. *El desafío del blanco: Goya y Esteve, retratistas de la casa de Osuna; A propósito de la donación de Alzaga*. Madrid: Museo Nacional del Prado, 2017.

Alderson, Brian. "New Playthings and Gigantick Histories: The Nonage of English Children's Books." *Princeton University Library Chronicle* 60, no. 2 (Winter 1999): 178–95.

Álvarez Barrientos, Joaquín. "Cultura y política entre siglos." In Álvarez Barrientos, *Se hicieron literatos para ser políticos*, 11–24.

Álvarez Barrientos, Joaquín. "El modelo femenino en la novela española del siglo XVIII." *Hispanic Review* 63, no. 1 (Winter 1995): 1–18.

Álvarez Barrientos, Joaquín. "Los hombres de letras." In Álvarez Barrientos, López, and Urzainqui, *La República de las letras*, 19–62.

Álvarez Barrientos, Joaquín, ed. *Se hicieron literatos para ser políticos: Cultura y política en la España de Carlos IV y Fernando VII*. Madrid: Servicio de Publicaciones de la Universidad de Cádiz, 2004.

Álvarez Barrientos, Joaquín, François López, and Inmaculada Urzainqui, eds. *La República de las letras en la España del siglo XVIII*. Madrid: Consejo Superior de Investigaciones Científicas, 1995.

Anes, Gonzalo. "La agricultura española desde comienzos del siglo XIX hasta 1868: Algunos problemas." In *Ensayos sobre la economía española a mediados del siglo XIX*, edited by Pedro Schwartz and Gabriel Tortella Casares, 235–63. Madrid: Ariel, 1970.

Añón Feliu, Carmen. *"El Capricho" de la Alameda de Osuna*. Madrid: Fundación Caja Madrid, 2001.

Antonelli, Francesca. "Scrittura, sociabilita e strategie di persuasione: Marie-Anne Paulze-Lavoisier, secrétaire (1758–1836)." PhD diss., EHESS-Università di Bologna, Paris, 2021.

Argemí, Lluís. "Agronomía y revolución agraria en España (1750–1820)." In *Agronomía y fisiocracia en España (1750–1820)*, edited by Ernest Lluch and Lluís Argemí, 1–43. Valencia: Institución Alfonso El Magnánimo, 1985.

Astigarraga, Jesús. "Economic Societies and the Politicisation of the Spanish Enlightenment." In *The Spanish Enlightenment Revisited*, edited by Jesús Astigarraga, 63–81. Oxford: Voltaire Foundation, 2015.

Astigarraga, Jesús. *Los Ilustrados vascos: Ideas, instituciones y reformas económicas en España*. Barcelona: Editorial Crítica, 2003.

Astigarraga, Jesús. *A Unifying Enlightenment: Institutions of Political Economy in Eighteenth-Century Spain (1700–1808)*. Leiden: Brill, 2020.

Atienza Hernández, Ignacio. *Aristocracia, poder y riqueza en la España Moderna: la Casa de Osuna, siglos XV–XIX*. Madrid: Siglo veintiuno de España, 1987.

Atienza Hernández, Ignacio. "Mujeres que mandan: aristócratas y ciclo vital en el siglo XVIII." In Deusa, *Historia de las mujeres*, 437–76.

Azcárate, Isabel. *Una niña regidora honoraria de la ciudad de Cádiz*. Cádiz: Quórum Libros, 2000.

Babilas, Dorota. "From Female Accomplishment to Botanical Science: Mary Delany's 'Paper Mosaicks'." *Literature Compass* 10, no. 8 (August 2013): 631–42.

Balland, Antoine. *La Chimie alimentaire dans l'œuvre de Parmentier*. París: J.-B. Baillière et fils, 1902.

Barrera-Osorio, Antonio. *Experiencing Nature: The Spanish American Empire and the Early Scientific Revolution*. Austin: University of Texas Press, 2006.

Barry, Jonathan, and Colin Jones. "Introduction." In Barry and Jones, *Medicine and Charity before the Welfare State*, 1–13.

Barry, Jonathan, and Jones Colin, eds. *Medicine and Charity before the Welfare State*. London: Routledge, 1991.

Batsaki, Yota, Sarah Burke Cahalan, and Anatole Tchikine, eds. *The Botany of Empire in the Long Eighteenth Century*. Washington, DC: Dumbarton Oaks Research Library and Collection, 2016.

Becker, Peter, and William Clark, eds. *Little Tools of Knowledge: Historical Essays on Academic and Bureaucratic Practices.* Ann Arbor: University of Michigan Press, 2001.

Bensaude-Vincent, Bernadette, and Christine Blondel, eds. *Science and Spectacle in the European Enlightenment.* Aldershot: Ashgate, 2007.

Bensaude-Vincent, Bernadette, and Ferdinando Abbri, eds. *Lavoisier in European Context.* Canton, MA: Science History, 1995.

Benzaquen, Adriana. "Childhood, Identity and Human Science in the Enlightenment." *History Workshop Journal* 57, no. 1 (Spring 2004): 35–57.

Beretta, Marco. "Italian Translations of the *Méthode de Nomenclature Chimique* and the *Traité Élémentaire de Chimie*: The case of Vicenzo Dandolo." In Bensaude-Vincent and Abbri, *Lavoisier in European Context*, 224–37.

Berg, Maxine, and Elizabeth Eger, eds. *Luxury in the Eighteenth Century: Debates, Desires and Delectable Goods.* Basingstoke: Palgrave, 2002.

Berman, Alex. "Parmentier, Antoine-Agustin." In *Dictionary of Scientific Biography*, ed. Charles C. Gillispie. Vol. 10. New York: Charles Scribner's Sons, 1980.

Bertomeu Sánchez, José Ramón. "The Colic of Madrid (1788–1814): Experts, Poisons, Politics, and War at the End of the Ancien Régime in Spain." *Social History of Medicine* 33, no. 3 (August 2020): 728–48.

Bertomeu Sánchez, José Ramón. "Pedro Gutiérrez Bueno (1745–1822) y las relaciones entre la quimica y la farmacia durante el último tercio del siglo XVIII." *Hispania* 61 (July 2010): 539–61.

Bertucci, Paola. *Artisanal Enlightenment: Science and the Mechanical Arts in Old Regime France.* New Haven: Yale University Press, 2017.

Bittel, Carla, Elaine Leong, and Christine von Oertzen, eds. *Working with Paper: Gendered Practices in the History of Knowledge.* Pittsburgh: University of Pittsburgh Press, 2019.

Blair, Ann. "Noël-Antoine Pluche as a Jansenist Natural Theologian." *Intellectual History Review* 26, no. 1 (January 2016): 91–99.

Bleichmar, Daniela. *Visible Empire: Botanical Expeditions and Visual Culture in the Hispanic Enlightenment.* Chicago: University of Chicago Press, 2012.

Block, Brian Philip. "Liberté, égalité, Parmentier! A Critical Study of the Life and Work of Antoine-Augustin Parmentier, Pharmacist Extraordinaire." PhD diss., Birkbeck College, University of London, 2009.

Bolufer, Mónica. "Transformaciones culturales: luces y sombras." In Deusa, *Historia de las mujeres*, 479–510.

Bolufer, Mónica, and Montserrat Cabré. "La Querelle des femmes en Espagne: Bilan sur l'histoire d'un débat." In Dubois-Nayt, Henneau, and von Kulessa, *Revisiter la "querelle des femmes"*, 31–45.

Bolufer Peruga, Mónica. "Actitudes y discursos sobre la maternidad en la España del siglo XVII: la cuestión de la lactancia." *Historia Social* 14 (1992): 3–22.

Bolufer Peruga, Mónica. "Conversations from a Distance: Spanish and French Eighteenth-Century Women Writers." In *A Companion to Spanish Women's Studies*, edited by Xon de Ros and Geraldine Hazbun, 175–188. Woodbridge: Tamesis, 2011.

Bolufer Peruga, Mónica. "Debate de los sexos y discurso de progreso en la Ilustración Española." In *Modernidad Iberoamericana: Cultura, política y cambio social*, edited by Francisco Colom González, 321–50. Madrid: Iberoamericana, 2009.

Bolufer Peruga, Mónica. "Desde la periferia: mujeres de la Ilustración en "Province." In *La*

Real Academia de Bellas Artes de San Carlos en la Valencia Ilustrada, edited by Romà Calle, 76–82. Valencia: Universitat de Valencia, 2009.

Bolufer Peruga, Mónica. "The Enlightenment in Spain: Classic and New Historiographical Perspectives." In Franklin Lewis, Bolufer Peruga, and Jaffe, *Routledge Companion*, 3–16.

Bolufer Peruga, Mónica. "Galerías de mujeres ilustres o el sinuoso camino de la excepción a la norma cotidiana (siglos XV–XVIII)." *Hispania* 60, no. 1 (2000): 181–224.

Bolufer Peruga, Mónica. "Gender and the Reasoning Mind: Introduction." In Knott and Taylor, *Women, Gender and Enlightenment*, 189–94.

Bolufer Peruga, Mónica. "Josefa Amar y Borbón (1749–1733): An Intellectual Woman." In *Women, Enlightenment and Catholicism: A Transnational Biographical History*, edited by Ulrich L. Lehner, 50–63. London: Routledge, 2018.

Bolufer Peruga, Mónica. "Las mujeres en la cultura de la Ilustración." In *Ilustración, ciencia y técnica en el siglo XVIII español*, edited by Enrique Martínez Ruiz y Magdalena de Pazzis and Pi Corrales, 209–32. Valencia: Universidad de Valencia, 2008.

Bolufer Peruga, Mónica. *Mujeres e Ilustración: La construcción de la feminidad en la Ilustración española*. Valencia: Institució Alfons el Magnànim, 1998.

Bolufer Peruga, Mónica. "Mujeres y hombres en los espacios del Reformismo Ilustrado: Debates y estrategia." *hMiC: història moderna i contemporània* 1 (2003): 155–70.

Bolufer Peruga, Mónica. "Neither Male, Nor Female: Rational Equality in the Early Spanish Enlightenment." In Knott and Taylor, *Women, Gender and Enlightenment*, 389–409.

Bolufer Peruga, Mónica. "New Inflections of a Long Polemic: The Debate between the Sexes in Enlightenment Spain." In *A New History of Iberian Feminisms*, edited by Silvia Bermúdez and Roberta Johnson, 38–49. Toronto: University of Toronto Press, 2018.

Bolufer Peruga, Mónica. "Pedagogía y moral en el siglo de las luces: Las escritoras francesas y su recepción en España." *Revista de historia moderna: Anales de la Universidad de Alicante* 20 (2002): 251–92.

Mónica Bolufer Peruga, "Women of Letters in Eighteenth-Century Spain: Between Tradition and Modernity." In Jaffe and Franklin Lewis, *Eve's Enlightenment*, 17–32.

Bolufer Peruga, Mónica. "Women in Patriotic Societies: A Spanish Debate in a European Context." In *Society Ladies and Philanthropy during the Spanish Enlightenment: La Junta de Damas de Honor y Mérito, 1787–1823*, edited by Catherine M. Jaffe and Elisa Martín-Valdepeñas Yagüe. Baton Rouge: Louisiana State University Press, forthcoming.

Bolufer, Mónica, and Elena Serrano. "Maritime Crossroads: The Knowledge Pursuits of María de Betancourt (Tenerife, 1758–1824) and Joana de Vigo (Menorca, 1779–1855)." In Maerker, Serrano, and Werrett, "Enlightened Female Networks."

Bolufer, Mónica, Isabel Morant Deusa, María José de la Pascua Sánchez, Gloria Espigado Tocino, Inmaculada Urzainqui Miqueleiz, and Juan Gomis Coloma, eds. *Mujeres y modernización: Estrategias culturales y prácticas sociales (siglos XVIII–XX)*. Madrid: Instituto de la Mujer, 2008.

Bonet, Antonio, and Beatriz Blasco, eds. *Fernando VI y Bárbara de Braganza: Un reinado bajo el signo de la paz (1746–1759)*. Madrid: Real Academia de Bellas Artes de San Fernando, 2002.

Bosch, Mineke. "Persona and the Performance of Identity: Parallel Developments in the Biographical Historiography of Science and Gender, and the Related Uses of Self Narrative." *L'homme: Zeitschrift für feministische Geschichtswissenschaft* 24, no. 2 (2013): 11–22.

Bottigheimer, Ruth B. "The Book on the Booksheller's Shelf and the Book in the English Child's Hand." In Ruwe, *Culturing the Child*, 3–28.

Bourde, André J. *Agronomie et agronomes en France au XVIIIe siècle*. Vol. 2. Paris: S.E.V.P.E.N, 1967.

Bravo-Villasante, Carmen. *Historia de la literatura infantil española*. Madrid: Doncel, 1972.

Bray, Francesca. *Technology, Gender and History in Imperial China: Great Transformations Reconsidered*. London: Routledge, 2013.

Bret, Patrice. "Un bateleur de la science: Le 'machiniste-physicien' François Bienvenu et la diffusion de Franklin et Lavoisier." *Annales Historiques de la Révolution Française* 338 (2004): 95–127.

Briggs, Julia. "'Delightful Task!' Women, Children and Reading in the Mid-Eighteenth Century." In Ruwe, *Culturing the Child*, 67–84.

Brokmann, Sophia. *The Science of Useful Nature in Central America: Landscapes, Networks and Practical Enlightenment, 1784–1838*. Cambridge, UK: Cambridge University Press, 2020.

Brooke, Hedley. "Natural Theology." In *Science and Religion: A Historical Introduction*, edited by Gary Ferngren, 163–75. Baltimore: John Hopkins University Press, 2002.

Brown, Penny. *A Critical History of French Children's Literature (1600–1830)*. London: Routledge, 2008.

Burton, June K. *Napoleon and the Woman Question: Discourse of the Other Sex in French Education, Medicine and the Medical Law 1779–1815*. Lubbock: Texas Tech University Press, 2007.

Cabré, Montserrat, and Teresa Ortiz, eds. *Sanadoras, matronas y médicas en Europa, siglos XII–XX*. Barcelona: Icaria, 2001.

Cabré Pairet, Montserrat, and Mónica Bolufer Peruga, eds. "La Querella de las Mujeres: Nuevas perspectivas historiográficas." *Arenal* 20 (2013).

Calcografía Nacional. *Antonio de Sancha, reinventor de lecturas y hacedor de libros (1720–1790)*. Madrid: Real Academia de Bellas Artes de San Fernando, 1997.

Calderón España, María Consolació. "Presencia de la mujer en las reales sociedades económicas de amigos del país (1775–1808)." *Foro de Educación* 12 (2010): 185–231.

Campbell Orr, Clarissa. "Aristocratic Feminism, the Learned Governess, and the Republic of Letters." In Knott and Taylor, *Women, Gender and Enlightenment*, 306–25.

Cañizares-Esguerra, Jorge. "Iberian Science in the Renaissance: Ignored How Much Longer?" *Perspectives on Science* 12, no. 1 (Spring 2004): 86–124.

Carbonell i Esteller, Montserrat. "Trabajo femenino y economías familiars." In Deusa, *Historia de las mujeres*, 2:237–62.

Carlyle, Margaret. "Breastpump Technology and 'Natural' Motherly Milk in Enlightenment France." *Women's Studies International Forum* 60 (January–February 2017): 89–96.

Carrasco, Paula. "Vida y obra de Pedro Gutiérrez Bueno." *Boletín de la Sociedad Española de Historia de la Farmacia* 15 (1964): 154–69; 16 (1965): 10–24, 71–86, 101–18, 153–77.

Carreras Panchón, Antonio. *El problema del niño expósito en la España Ilustrada*. Salamanca: Universidad de Salamanca, 1977.

Carreras Panchón, Antonio. "La mortalidad en las inclusas españolas (1800–1808)." *Cuadernos de Historia de la Medicina Española* 14 (1975): 261–68.

Cartaña i Pinén, Jordi. "La agronomía en la España del setecientos." In *Técnica e ingeniería en España (III): El siglo de las luces. De la industria al ámbito agroforestal*, edited by Manuel Silva Suárez, 409–52. Zaragoza: Prensas Universitarias de Zaragoza, 2005.

Cavallo, Sandra. "The Motivations of Benefactors: An Overview of Approaches to the Study of Charity." In Barry and Jones, *Medicine and Charity*, 46–62.

Cavazza, Marta. "Between Modesty and Spectacle: Women and Science in Eighteenth-Century Italy." In *Italy's Eighteenth Century: Gender and Culture in the Age of the Grand Tour*, edited by Paula Findlen, Wendy Wassyng Roworth, and Catherine M. Sama, 275–302. Stanford, CA: Stanford University Press, 2009.

Cavazza, Marta. *Laura Bassi. Donne, genere e scienza nell'Italia del Settecento*. Milan: Editrice Bibliografica, 2020.

Charles, Loïc, and Christine Théré. "Les femmes économistes: The Place of Women in the Physiocratic Community." In Maerker, Serrano, and Werrett, "Enlightened Female Networks."

Chappey, Jean-Luc. *La Société des observateurs de l'homme (1799–1804): Des anthropologues au temps de Bonaparte*. Paris: Société des études robespierristes, 2002.

Chivelet, Mercedes. *La prensa infantil en España, desde el siglo XVIII hasta nuestros días*. Madrid: Editorial SM, 2009.

Clark, Peter. *British Clubs and Societies 1580–1800: The Origins of an Associational World*. Oxford: Oxford University Press, 2002.

Clarke, Norma: "The Cursed Barbault Crew: Women Writers and Writing for Children in the Late Eighteenth Century." In *Opening the Nursery Door: Reading, Writing and Childhood (1600–1900)*, edited by Mary Hilton, Morang Styles, and Victor Watson, 91–103. London: Routledge, 1997.

Clarke, Norma. *The Rise and Fall of the Woman of Letters*. London: Pimlico, 2003.

Clarke, William, Jan Golinski, and Simon Schaffer, eds. *The Sciences in Enlightened Europe*. Chicago: University of Chicago Press, 1999.

Colley, Linda. *Britons: Forging the Nation 1707–1837*. New Haven: Yale University Press 2005.

Comín, Fernando. "Canga Argüelles: Un planteamiento realista de la Hacienda liberal." In Fuentes Quintana, *Economía y economistas españoles*, 4:413–39.

Cotarelo y Mori, Emilio. *Biografía de D. Antonio de Sancha: Un gran editor español del siglo XVIII*. Madrid: Cámaras Oficiales del Libro de Madrid y Barcelona, 1924.

Cristiani, Andrea. "Dall'entusiasmo al plagio: Ossidazioni e riduzioni vervali nella chimica divulgativa di Giusseppe Compagnoni." In Medri, *Giusseppe Compagnoni*.

Dabhoiwala, Faramerz. *The Origins of Sex: A History of the First Sexual Revolution*. London: Penguin, 2013.

Daston, Lorraine. "Afterwards: The Ethos of Enlightenment." In Clarke, Golinski, and Schaffer, *Sciences in Enlightened Europe*, 495–504.

Daston, Lorraine. "The Naturalized Female Intellect." *Science in Context* 5, no. 2 (Autumn 1992): 209–35.

Daston, Lorraine, and Gianna Pomata. "The Faces of Nature: 'Visibility and Authority." In Daston and Pomata, *Faces of Nature*, 1–16.

Daston, Lorraine, and Gianna Pomata, eds. *The Faces of Nature in Enlightenment Europe*. Berlin: Berliner Wissenschafts-verlag, 2003.

Daston, Lorraine, and H. Otto Sibum. "Introduction: Scientific Personae and Their Histories." *Science in Context* 16, nos. 1–2 (March 2003): 1–8.

Daston, Lorraine, and Katharine Park, eds. *The Cambridge History of Science*, vol. 3, *Early Modern Science*. New York: Cambridge University Press, 2006.

Davids, Karel. "Gatekeeping. Who Defined 'Useful Knowledge' in Early Modern Times?" In *History of Technology*, edited by Ian Inkster, 31:69–88. London: Bloomsbury Academic, 2012.

De Baere, Benoît. *Trois introductions à l'Abbé Pluche: Sa vie, son monde, ses livres*. Geneva: Droz, 2001.

Demerson, Paula. "Catálogo de las socias de honor y mérito de la Junta de damas matritense (1787–1811)." *Anales del Instituto de Estudios Madrileños* 7 (1971): 269–74.

Demerson, Paula. *Esbozo de Biblioteca de la Juventud Ilustrada (1740–1808)*. Oviedo: Cátedra Feijoo, 1976.

Demerson, Paula. "La Real Inclusa de Madrid a finales del siglo XVIII." *Anales del Instituto de Estudios Madrileños* 8 (1972): 261–72.

Demerson, Paula. *María Francisca de Sales y Portocarreño, Condesa del Montijo: Una figura de la Ilustración*. Madrid: Editora Nacional, 1975.

Demerson, Paula, and Jorge Demerson. "La Sociedad Económica de Amigos del País de Ciudad Rodrigo." *Cuadernos de historia moderna y contemporánea* 3 (1982): 35–60.

Deusa, Isabel Morant, ed. *Historia de las mujeres en España y América Latina*. Madrid: Cátedra, 2005.

Donato, Clorinda, and Ricardo López. *Enlightenment Spain and the "Encyclopédie Méthodique."* Oxford: Oxford University Studies in the Enlightenment, 2015.

Donato, Clorinda, and Manuel Romero. "'Todos los progresos que ha hecho el entendimiento humano': Knowledge, Networking and the Encyclopedic Turn in Enlightenment Spain." In Franklin Lewis, Bolufer Peruga, and Jaffe, *Routledge Companion*, 271–85.

Doyle, Nora. *Maternal Bodies: Redefining Motherhood in Early America*. Chapel Hill: University of North Carolina Press, 2018.

Drake, T. G. H. "Infant Feeding in England and in France from 1750 to 1800." *American Journal of Diseases of Children* (1930): 1049–61.

Driel, Joppe van. "Ashes to Ashes: The Stewardship of Waste and Oeconomic Cycles of Agricultural and Industrial Improvement, 1750–1800." *History and Technology* 30, no. 3 (2014): 177–206.

Dubois-Nayt, Armel, Marie-Elisabeth Henneau, and Rotraud von Kulessa, eds. *Revisiter la "querelle des femmes": Discours sur l'egalité/inégalité des sexes en Europe, de 1400 aux lendemains de la Révolution*. Saint-Étienne: Publications de l'Université de Saint-Étienne, 2015.

Earle, Rebecca. "Food, Colonialism and the Quantum of Happiness." *History Workshop Journal* 84 (Autum 2017): 170–93.

Earle, Rebecca. "Potatoes and the Hispanic Enlightenment." *Americas* 75, no. 4 (Winter 2018): 639–60.

Earle, Rebecca. "Promoting Potatoes in Eighteenth-Century Europe." *Eighteenth-Century Studies* 51, no. 2 (Winter 2017): 147–62.

Enciso Recio, Luis Miguel. *Las sociedades económicas en el siglo de las luces*. Madrid: Real Academia de la Historia, 2010.

Enciso Recio, Luis Miguel. "Presencia y actividad de los burgueses en las Sociedades Económicas." *Cuadernos de estudios del siglo XVIII* 3–4 (1993–1994): 3–60.

Equipo Madrid. *Carlos III, Madrid y la Ilustración: Contradicciones de un proyecto reformista*. Madrid: Siglo XXI, 1988.

Enterría, Josefa Gómez de. *Voces de la economía y el comercio en el español del siglo XVIII*. Alcalá: Universidad de Alcalá, 1996.

Escavy Zamora, Ricardo. "Estudio introductorio." In *Diego Clemencín. Lecciones de Gramática y Ortografía Castellana. Edición Facsimilar de la edición de Madrid. Imprenta de D. Miguel de Burgos, 1842*, 11–88. Murcia: Universidad de Murcia. Servicio de Publicaciones, 2012.

Espina Pérez, Pedro. *Historia de la Inclusa de Madrid.* Madrid: Defensor del Menor en la Comunidad de Madrid, 2005.

Feijoo, Benito J. *Cartas Eruditas y Curiosas.* 5 vols. Madrid: publishing house unknown, 1742–1760.

Feingold, Mordechai, and Victor Navarro-Brotons, eds. *Universities and Science in the Early Modern Period.* Dordrecht: Springer, 2006.

Felten, Sebastian, and Christine von Oertzen. "Bureaucracy as Knowledge." *Journal for the History of Knowledge* 1 (2020): 1–8.

Fernández Pérez, Joaquín. "La ciencia ilustrada y las sociedades económicas de amigos del país." In *Carlos III y la ciencia de la Ilustración*, edited by Manuel Selles, José Luis Peset, and Antonio Lafuente, 247–45. Madrid: Alianza Universidad, 1988.

Fernández Quintanilla, Paloma. *La IX Duquesa de Osuna: Una Ilustrada en la corte de Carlos III.* Madrid: Ediciones 12 Calles, 2017.

Fernández-Quintanilla, Paloma. *La mujer ilustrada en la España del siglo XVIII.* Madrid: Dirección General de Juventud y promoción Sociocultural, 1981.

Fernández Vargas, Valentina. "Informe sobre el archivo de la Antigua Inclusa de Madrid, hoy Instituto Provincial de Puericultura." In Espina Pérez, *Historia de la Inclusa de Madrid*, 47–57.

Fernández Vargas, Valentina, ed. *El Madrid de las mujeres: Aproximación a una presencia invisible (1561–1833).* Madrid: Dirección General de la Mujer, 2007.

Figueroa, Carmen Yebes Muñoz de. *La Condesa-Duquesa de Benavente: una vida en unas cartas.* Madrid: Espasa-Calpe, 1955.

Fildes, Valerie. *Breast, Bottles and Babies: A History of Infant Feeding.* Edinburgh: Edinburgh University Press, 1986.

Fildes, Valerie. *Wet Nursing: A History from Antiquity to the Present.* New York: Basil Blackwell, 1988.

Findlen, Paula. "Becoming a Scientist: Gender and Knowledge in Eighteenth-Century Italy." *Science in Context* 16, nos. 1–2 (March 2003): 59–87.

Findlen, Paula. "A Forgotten Newtonian: Women and Science in the Italian Provinces." In Clark, Golinski, and Schaffer, *Sciences in Enlightened Europe*, 313–49.

Findlen, Paula. "The Scientist's Body: The Nature of a Woman Philosopher in Enlightenment Italy." In Daston and Pomata, *Faces of Nature*, 211–36.

Findlen, Paula. "Translating the New Science: Women and the Circulation of Knowledge in Enlightenment Italy." *Configurations* 3, no. 2 (Spring 1995): 167–206.

Findlen, Paula. "Women on the Verge of Science: Aristocratic Women and Knowledge in Early Eighteenth-Century Italy." In Knott and Taylor, *Women, Gender and Enlightenment*, 265–87.

Fonseca, Palmira. *Un hacendista asturiano: José Canga Argüelles.* Oviedo: Real Instituto de Estudios Asturianos, 1995.

Fontana, Joseph. *La Hacienda en la historia de España 1700–1931.* Madrid: Instituto de Estudios Fiscales, 1980.

Fontes da Costa, Palmira. "Women and the Popularisation of Botany in Early Nineteenth-Century Portugal: The Marquesa de Alorna's *Botanical Recreations*." In Papanelopoulou, Nieto-Galán, and Perdiguero, *Popularizing Science and Technology*, 43–63.

Forniés Casals, José Francisco. *La Real Sociedad Económica Aragonesa de Amigos del País.* Zaragoza: Caja de Ahorros de la Inmaculada, 2000.

Forrest, Alan. "Poverty." In *The Oxford Handbook of the Ancient Régime*, edited by William Doyle, 167–82. Oxford: Oxford University Press, 2012.

Fox, Robert, and Augustí Nieto-Galán, eds. *Natural Dyestuffs and Industrial Culture in Europe.* Canton, MA: Science History, 1999.

Franco Rubio, Gloria. "El talento no tiene sexo: debates sobre la educación femenina en la España moderna." In *El alma de las mujeres: ámbitos de espiritualidad femenina en la modernidad (siglos XVI–XVIII)*, edited by Javier Burrieza Sánchez, 363–93. Valladolid: Universidad de Valladolid, 2015.

Franco Rubio, Gloria Ángeles. "Captar súbditos y crear ciudadanos, doble objetivo de los "Amigos del País" en el siglo XVIII." *Historia Social* 64 (2009): 3–23.

Franco Rubio, Gloria Ángeles. "Las Sociedades Económicas de Amigos del País: Un observatorio privilegiado para la práctica política y el nacimiento de la ciudadanía en el Antiguo Régimen." In *Ilustración, ilustraciones*, edited by Jesús Astigarraga, M. Victoria López-Cordón, and José M. Urkia, 1:351–68. Donostia: Real Sociedad Bascongada de los Amigos del País-Sociedad Estatal de Conmemoraciones Culturales, 2011.

Franklin Lewis, Elizabeth. "Actos de caridad: Women's Charitable Work in Eighteenth-Century Spain." *Dieciocho: Hispanic Enlightenment* 31 (2008): 267–68.

Franklin Lewis, Elizabeth. "Feijoo, Josefa Amar y Borbon and the Feminist Debate in Eighteenth-Century Spain." *Dieciocho: Hispanic Enlightenment* 12 (1989): 188–203.

Franklin Lewis, Elizabeth. "Women as Public Intellectuals during the Hispanic Enlightenment: The Case of Josefa Amar y Borbón's *Ensayo histórico-apologético de la literature Española*." In Franklin Lewis, Bolufer Peruga, and Jaffe, *Routledge Companion*, 112–25.

Franklin Lewis, Elizabeth. *Women Writers in the Spanish Enlightenment: The Pursuit of Happiness.* Aldershot: Ashgate, 2004.

Franklin Lewis, Elizabeth, Mónica Bolufer Peruga, and Catherine Jaffe, eds. *Routledge Companion to the Hispanic Enligthenment.* London: Routledge, 2019.

Fuchs, Rachel G. *Abandoned Children: Foundlings and Child Welfare in Nineteenth-Century France.* Albany: State University of New York Press, 1984.

Fuentes Quintana, Enrique, ed. *Economía y economistas españoles: la Ilustración.* Barcelona: Galaxia Gutenberg, 2000.

Fyfe, Aileen. "Introduction." In *Science for Children*, 7 vols., edited by Aileen Fyfe, 1:xi-xxii. Bristol: Thoemmes Press, 2003.

Fyfe, Aileen. "Reading Children's Books in Late Eighteenth-Century Dissenting Families." *Historical Journal* 43, no. 2 (June 2000): 453–73.

Fyfe, Aileen, and Bernard Lightman, eds. *Science in the Marketplace: Nineteenth-Century Sites and Experiences.* Chicago: University of Chicago Press, 2007.

Gago, Ramón. "The New Chemistry in Spain." *Osiris* 4 (1988): 169–92.

García Ballester, Luis, ed. *Historia de la ciencia y de la técnica en la Corona de Castilla.* 4 vols. Valladolid: Junta de Castilla y León, 2002.

García Belmar, Antonio, and José Ramón Bertomeu Sánchez. "Pedro Gutiérrez Bueno (1745–1822), los libros de texto y los nuevos públicos de la química en el último tercio del siglo XVIII." *Dynamis: Acta Hispanica ad Medicinae Scientiarumque Historiam Illustrandam* 21 (2001): 351–74.

García Belmar, Antonio, and José Ramon Bertomeu Sánchez. "Spanish Chemistry Textbooks during the Late Eighteenth Century: Building Up a New Genre of Scientific Literature." In Feingold and Navarro-Brotons, *Universities and Science in the Early Modern Period*, 241–57.

García Padrino, Jaime. *Así pasaron muchos años: En torno a la literatura infantil española.* Cuenca: Ediciones de la Universidad de Castilla-La Mancha, 2001.

Gavroglu, Kostas, Manolis Patiniotis, Faidra Papanelopoulou, Ana Simões, Ana Carneiro, Maria Paula Diogo, José Ramón Bertomeu Sánchez, Antonio García Belmar, and Agustí Nieto-Galan. "Science and Technology in the European Periphery: Some Historiographical Reflections." *History of Science* 46, no. 2 (June 2008): 154–75.

Gevrey, Françoise, Julie Boch, and Jean-Louis Haquette, eds. *Écrire la nature au XVIIIe siècle: Autour de l'abbé Pluche*. Paris: PUPS, 2006.

Gil, Alberto, and Ana Boned, eds. *Diccionario biográfico del trienio liberal*. Madrid: El Museo Universal, 1991.

Gillispie, Charles C. *Science and Polity in France at the End of the Old Regime*. Princeton, NJ: Princeton Unversity Press, 1980.

Ginzburg, Lori D. *Women and the Work of Benevolence: Morality, Politics, and Class in the Nineteenth-Century United States*. New Haven: Yale University Press, 1990.

Glénisson, Jean. "Le livre pour la jeunesse." In *Histoire de l'edition française*, edited by Henri-Jean Martin and Roger Chartier, 3:417–37. Paris: Promodis, 1986.

Glénisson, Jean, and Ségolène Le Men, eds. *Le Livre d'enfance et de jeunesse en France*. Bordeaux: Société des Bibliophiles de Guyenne, 1994.

Goddard, Nicholas. "Agricultural Literature and Societies." In *The Agrarian History of England and Wales 1750–1850*, edited by G. E. Mingay, 6:361–83. Cambridge, UK: Cambridge University Press, 1989.

Golinski, Jan. "Science in the Enlightenment, Revisited." *History of Science* 49 no. 2 (June 2011): 217–31.

González, Antonio. "El real Gabinete de Historia Natural." In *Madrid, ciencia y corte*, 247–51. Madrid: Gráfica Futura, 1999.

Granjel, Luis. *Historia de la Real Academia Nacional de Medicina*. Madrid: Taravilla, 2006.

Grenby, M. O. *The Child Reader: 1700–1840*. Cambridge, UK: Cambridge University Press, 2011.

Grenby, M. O. "The Origins of Children's Literature." In Grenby and Immel, *Cambrige Companion*, 3–18.

Grenby, M. O., and Andrea Immel, eds. *The Cambridge Companion to Children's Literature*. Cambridge, UK: Cambridge University Press, 2009.

Groebner, Valentin. *Who Are You? Identification, Deception, and Surveillance in Early Modern Europe*. Translated by Mark Kyburz and John Peck. New York: Zone Books, 2007.

Guerra, Corinna. "Chimica per la donne." In *Scorci di storia della scienza*, edited by Frank Martin L. and C. Pogliano, 127–39. Pisa: Pisa University Press, 2010.

Haidt, Rebecca. "The Wife, the Maid, and the Woman in the Street." In Jaffe and Franklin Lewis, *Eve's Enlightenment*, 115–27.

Haidt, Rebecca. *Women, Work and Clothing in Eighteenth-Century Spain*. Oxford: Voltaire Foundation, 2011.

Hardwick, Juli. *Sex in an Old Regime City: Young Workers and Intimacy in France, 1660–1789*. Oxford: Oxford University Press, 2020.

Harvey, Karen. *The Little Republic: Masculinity and Domestic Authority in Eighteenth-Century Britain*. Oxford: Oxford University Press, 2012.

Havelange, Isabelle, and Ségolène Le Men. *Le Magasin des enfants: La littérature pour la jeunesse (1750–1830)*. Montreuil: Bibliothèque Robert Desnos, 1988.

Heesen, Anke te. "Accounting for the Natural World: Double-Entry Bookkeeping in the Field." In Schiebinger and Swan, *Colonial Botany*, 237–51.

Heesen, Anke te. "The Notebook: A Paper-Technology." In *Making Things Public: Atmospheres of Democracy*, edited by Bruno Latour and Peter Weibel, 582–89. Cambridge, MA: MIT Press, 2005.

Hernández González, Miguel. *Opúsculos científicos de José Viera y Clavijo*. La Orotava: Fundación Canaria Orotava de Historia de la Ciencia, 2012.

Herr, Richard. *The Eighteenth-Century Revolution in Spain*. Princeton, NJ: Princeton University Press, 1958.

Hesse, Carla. "Women Intellectuals in the Enlightened Republic of Letters: Introduction." In Knott and Taylor, *Women, Gender and Enlightenment*, 259–63.

Hudson, Kenneth. *Patriotism with Profit: British Agricultural Societies in the Eighteenth and Nineteenth Centuries*. London: Hugh Evelyn, 1972.

Hufton, Olwen H. *The Poor of Eighteenth-Century France 1750–1789*. Oxford: Oxford University Press, 1974.

Hutton, Sarah. "The *Persona* of the Woman Philosopher in Eighteenth-Century England: Catharine Macaulay, Mary Hays, and Elizabeth Hamilton." *Intellectual History Review* 18, no. 3 (January 2008): 403–12.

Iglesias, María del Carmen. "La nueva sociabilidad: Mujeres nobles y salones literarios y políticos." In *Nobleza y sociedad en la España moderna*, edited by María del Carmen Iglesias, 2:179–230. Oviedo: Nóbel-Fundación Central Hispano, 1997.

Immel, Andrea. "Children's Books and Constructions of Childhood." In Grenby and Immel, *Cambrige Companion*, 19–34.

Immel, Andrea, and Michel Witmore. "Little Differences: Children, Their Books, and Culture in the Study of Early Modern Europe." In *Childhood and Children's Books*, 1–20.

Immel, Andrea, and Michael Witmore, eds. *Childhood and Children's Books in Early Modern Europe (1550–1800)*. London: Routledge, 2006.

Infantes, Víctor, François López, and Jean-François Botrel, eds. *Historia de la edición y de la lectura en España: 1472–1914*. Madrid: Fundación Germán Sánchez Ruipérez, 2003.

Jackson, Mary. V. *Engines of Instruction, Mischief and Magic: Children's Literature in England from Its Beginning to 1839*. Lincoln: University of Nebraska Press, 1989.

Jacobs, Margaret, and Dorothée Sturkenboom. "A Women's Scientific Society in the West: The Late Eighteenth-Century Assimilation of Science." *Isis* 94, no. 2 (June 2003): 217–52.

Jaffe, Catherine M. "Contesting the Grounds for Feminism in the Hispanic Eighteenth-Century: The Enlightenment and Its Legacy." In Franklin Lewis, Bolufer Peruga, and Jaffe, *Routledge Companion*, 69–82.

Jaffe, Catherine M. "Doña Leonara's Library: Women's Readings from *The Spectator* (1711) to *El Semanario de Salamanca* (1795)." In Jaffe and Franklin Lewis, *Eve's Enlightenment*, 178–96.

Jaffe, Catherine M., Elisa Martín-Valdepeñas Yagüe. *María Lorenza de los Ríos, marquesa de Fuerte-Híjar: vida y obra de una escritora del Siglo de las Luces*. Madrid: Iberoamericana-Vervuert, 2019.

Jaffe, Catherine M., and Elisa Martín-Valdepeñas Yagüe. "Sociabilidad, filantropía y escritura: María Lorenza de los Ríos y Loyo, Marquesa de Fuerte-Híjar (1761–1821)." In *Mujeres y culturas políticas en España, 1808–1845*, edited by Ana Yetana Laguna, 87–128. Barcelona: Universitat Autónoma de Barcelona, 2013.

Jaffe, Catherine M., and Elizabeth Franklin Lewis, eds. *Eve's Enlightenment: Women's Experiences in Spain and Spanish America, 1726–1839*. Baton Rouge: Louisiana State University Press, 2009.

Jardine, Nicholas, James Secord, and Emma C. Spary, eds. *The Cultures of Natural History.* Cambridge, UK: Cambridge University Press, 1996.

Jasanoff, Sheila. "Future Imperfect: Science, Technology, and the Imaginations of Modernity." In *Dreamscapes of Modernity: Sociotechnical Imaginaries and the Fabrication of Power,* edited by Sheila Jasanoff and Sang-Hyun Kim, 1–49. Chicago: University of Chicago Press, 2015.

Johnson, Eric J. "Chronology." In Grenby and Immel, *Cambrige Companion,* xiii–xvii.

Jones, Colin. *Charity and Bienfaisance: The Treatment of the Poor in the Montpellier Region 1740–1815.* Cambridge, UK: Cambridge University Press, 1982.

Jones, Peter M. *Agricultural Enlightenment: Knowledge, Technology, and Nature 1750–1840.* Oxford: Oxford University Press, 2016.

Jorland, Gérard, Annick Opinel, and Geroge Weisz, eds. *Body Counts: Medical Quantification in Historical and Sociological Perspective.* London: McGill-Queen's University Press, 2005.

Kaftka, Ben. "Hunting the Plumed Mammal: The History of 'Bureaucracy' in France, 1750–1850." In *Figures of Authority: Contributions towards a Cultural History of Governance,* edited by Peter Becker and Rüdiger von Krosigk, 119–26. Brussels: Peter Lang, 2008.

Kaftka, Ben. "Paperwork: The State of the Discipline." *Book History* 12 (2009): 340–53.

Keene, Melanie. "From Candles to Cabinets: 'Familiar Chemistry' in Early Victorian Britain." *Ambix* 60 (2013): 54–77.

Keene, Melanie. "Playing among the Stars: *Science in Sport, or the Pleasures of Astronomy* (1804)." *History of Education: Sources and Interpretations* 40 (2011): 521–42.

Keene, Melanie. *Science in Wonderland: The Scientific Fairy Tales of Victorian Britain.* Oxford: Oxford University Press, 2015.

Kitts, Sally A. *The Debate on the Nature, Role and Influence of Women in Eighteenth-Century Spain.* Lewiston, NY: Edwin Mellen Press, 1995.

Klein, Ursula, and Emma Spary, eds. *Materials and Expertise in Early Modern Europe: Between Market and Laboratory.* Chicago: University of Chicago Press, 2010.

Klein, Ursula, and Emma C. Spary. "Introduction: Why Materials?" In Klein and Spary, *Materials and Expertise,* 1–24.

Kohlstedt, Sally Gregory, and Helen Longino. "The Women, Gender and Science Question: What Do Research on Women in Science and Research on Gender and Science Have to Do with Each Other?" In "Women, Gender, and Scince: New Directions." Special issue, *Osiris* 12 (1997): 3–15.

Knoepflmacher, U. C. "Children's Texts and the Grown-up Reader." In Grenby and Immel, *Cambridge Companion,* 159–73.

Knott, Sara, and Barbara Taylor, eds. *Women, Gender and Enlightenment.* London: Palgrave Macmillan, 2005.

Koepp, Cynthia. "Curiosity, Science, and Experiential Learning in the Eighteenth Century: Reading the *Spectacle de la nature.*" In Immel and Witmore, *Childhood and Children's Books,* 153–80.

Koerner, Lisbet. *Linnaeus: Nature and Nation.* Cambridge, MA: Harvard University Press, 1999.

Koerner, Lisbet. "Women and Utility in Enlightenment Science." In "Gender and Early Modern Science Cluster," edited by Mary Terrall. Special issue, *Configurations* 3, no. 2 (Spring 1995): 233–55.

Kühnel, Florian. "'Minister-Like Cleverness, Understanding, and Influence of Affairs': Ambassadresses in Everyday Business and Courtly Ceremonies at the Turn of the Eighteenth Century." In *Practices of Diplomacy in Early Modern World, c. 1410–1800*, edited by Tracey A. Sowerby and Jan Hennings, 130–42. London: Routledge, 2017.

La Parra, Emilio. *La alianza de Godoy con los revolucionarios: España y Francia a fines del siglo XVIII*. Madrid: CSIC, 1992.

La Vopa, Anthony J. *The Labor of the Mind: Intellect and Gender in Enlightenment Cultures*. Philadelphia: University of Pennsylvania Press, 2017.

Labrador, Carmen, and J. Carlos de Pablos. *La Educación en los papeles periódicos de la Ilustración Española*. Madrid: Ministerio de Educación y Ciencia, 1989.

Lafuente, Antonio. *Guía del Madrid científico: ciencia y corte*. Madrid: CSIC, 1998.

Lafuente, Antonio, and Juan Pimentel. "La construcción de un espacio público para la ciencia: Escrituras y escenarios en la Ilustración española." In García Ballester, *Historia de la ciencia*, 4:111–56.

Lafuente, Antonio, and Nuria Valverde. *Los mundos de la ciencia en la Ilustración española*. Madrid: Fundación Española para la Ciencia y la Tecnología, 2003.

Larriba, Elisabel. "L'art au service de la divulgation scientifique: Le rôle des gravures dans le *Semanario de Agricultura y Artes dirigido a los párrocos*" (1797–1808)." *El argonauta español* 2 (2005). http://argonauta.imageson.org/document57.html.

Larriba, Elisabel. *Le public de la presse en Espagne à la fin du XVIIIe siècle (1781–1808)*. Paris: Honoré Champion, 1998.

Larriba, Elisabel. "Un instrument de la politique agraire de Godoy: Le Semanario de Agricultura y Artes dirigido a los Párrocos (1797–1808)." *Bulletin Hispanique* 104 (2002): 243–61.

Larriba, Elisabel. "Un intento de reforma agraria por y para las clases productoras: El Semanario de Agricultura y Artes dirigido a los párrocos." *Brocar: Cuadernos de Investigación Histórica* 23 (1999): 87–117.

Larriba, Elisabel, and Gérard Dufour. "Introducción." In *El Semanario de agricultura y artes dirigido a los párrocos (1797–1808)*, 9–62. Valladolid: Editorial Ámbito, 1997.

Lémonon Waxin, Isabelle. "La Savante des Lumières française, histoire d'une persona: Pratiques, représentations, espaces et réseaux." PhD diss., EHESS, 2019.

Leong, Elaine. *Recipes and Everyday Knowledge: Medicine, Science and the Household in Early Modern England*. Chicago: University of Chicago Press, 2018.

Leong, Elaine, and Alisha Rankin. "Testing Drugs and Trying Cures: Experiment and Medicine in Medieval and Early Modern Europe." In "Testing Drugs and Trying Cures." Special issue, *Bulletin of the History of Medicine* 91, no. 2 (Summer 2017): 157–82.

Leong, Elaine, and Alisha Rankin, eds. *Secrets and Knowledge in Medicine and Science, 1500–1800*. Farnham: Ashgate, 2011.

Lerman, Nina E. "Categories of Difference, Categories of Power: Bringing Gender and Race to the History of Technology." *Technology and Culture* 51, no. 4 (October 2010): 893–918.

Lerman, Nina E., Ruth Oldenziel, and Arwen P. Mohun, eds. *Gender and Technology: A Reader*. Baltimore: Johns Hopkins University Press, 2003.

Levene, Alysa. *Childcare, Health and Mortality at the London Foundling Hospital, 1741–1800: "Left to the Mercy of the World."* Manchester: Manchester University Press, 2007.

Limia y Menéndez, José Varela de. *Los directores de la Real Sociedad Económica Matritense de Amigos del País y las presidentas de su Junta de Damas de Honor y Mérito*. Madrid: Talleres de "El Eco Franciscano," 1925.

Llombart Rosa, Vicente A., and Jesús Astigarraga Goenaga. "Las primeras 'antorchas de la economía': Las sociedades económicas de amigos del país en el siglo XVIII." In Fuentes Quintana, *Economía y economistas españoles*, 3:677–707.

Llorens, Vicente. *Liberales y Románticos: Una emigración española en Inglaterra (1823–1834)*. Madrid: Editorial Castalia, 1968.

Lloyd, Harriet Olivia. "Rulers of Opinion: Women at the Royal Institution of Great Britain, 1799–1812." PhD diss., University College London, 2019.

Lloyd Hart, V. E. *John Wilkes and the Foundling Hospital at Aylesbury 1759–1768*. Aylesbury: HM & M, 1979.Lluch, Ernest, and Salvador Almenar Palau. "Difusión e influencia de los economistas clásicos en España (1776–1870)." In Fuentes Quintana, *Economía y economistas españoles*, 4:93–170.

López, Antonio, and Carmen Manso. *Cartografía del siglo XVIII: Tomás López en la Real Academia de la Historia*. Madrid: Real Academia de la Historia, 2006.

López, François. "El libro y su mundo." In Álvarez Barrientos, López, and Urzainqui, *La República de las letras*, 63–124.

López Cordón, Maria Victoria. "Traducciones y traductoras en la España de finales del siglo XVIII." In *Entre la marginación y el desarrollo: Mujeres y hombres en la historia. Homenaje a María Carmen García Nieto*, edited by Cristina Segura y Gloria Nielfa, 89–112. Madrid: Ediciones del Orto, 1996.

López-Cordón Cortezo, María Victoria. *Condición femenina y razón ilustrada: Josefa Amar y Borbón*. Zaragoza: Prensas Universitarias, 2005.

López-Cordón Cortezo, María Victoria. "The Enlightenment and Its Interpreters: Nobility, Bureaucrats, and Publicists." In Franklin Lewis, Bolufer Peruga, and Jaffe, *Routledge Companion*, 203–17.

López-Cordón Cortezo, María Victoria. "Josefa Amar y Borbón (1749–1833): Los debates ilustrados sobre las mujeres." In *Feminismos: Contribuciones desde la historia*, edited by Ángela En Cenarro and Régine Illion, 51–80. Zaragoza: Prensas Universitarias de Zaragoza, 2014.

López Ruiz, Antonio, and Eusebio Aranda-Muñoz. *Don Diego Clemencín (1765–1834): Ensayo bio-bibliográfico*. Murcia: Publicaciones Universidad de Murcia, 1948.

Loqueneux, Robert. "L'abbé Pluche, ou l'accord de la foi et de la raison à l'aube des Lumières." *Sciences et Techniques en Perspective* 2 (1998): 235–88.

Lowood, Henry E. *Patriotism, Profit and the Promotion of Science in the German Enlightenment: The Economic and Scientific Societies, 1760–1815*. New York: Garland, 1991.

Luna, Juan J. "Retrato de los duques de Osuna y sus hijos." In *Realidad e imagen: Goya, 1746–1828*, edited by F. Torralba, 100. Zaragoza: Gobierno de Zaragoza, 1996.

Madulid, Domingo A. "The Life and Work of Luis Née, Botanist of the Malaspina Expedition." *Archives of Natural History* 16 (1989): 33–48.

Maerker, Anna. "Political Order and the Ambivalence of Expertise: Count Rumford and Welfare Reform in Late Eighteenth-Century Munich." In "Expertise and the Early Modern State," edited by Eric H. Ash. Special issue, *Osiris* 25, no. 1 (2010): 213–30.

Maerker, Anna, Elena Serrano, and Simon Werrett, eds. "Enlightened Female Networks:
Gendered Ways of Producing Knowledge (1720–1830)." Special issue, *Notes and Records: The Royal Society Journal of the History of Science*. Forthcoming.

Malet, Antoni. "Newton in the Iberian Peninsula." In *The Reception of Isaac Newton in the Europe*, 3 vols., edited by S. Mandelbrote and H. Pulte, vol. 1, ch 6. London: Continuum, 2019.

Mander, Jeny. "No Woman Is an Island: The Female Figure in French Enlightenment Anthropology." In Knott and Taylor, *Women, Gender and Enlightenment*, 97–116.

Manson, Michel. *Les livres pour l'enfance et la jeunesse sous la Révolution*. Montreuil: INRP Bibliothèque Robert-Desnos, 1988.

Martin, Christophe. "Women's Curiosity and Its Double at the Dawn of the Enlightenment." In *Women and Curiosity in Early Modern England and France*, edited by Line Cottegnies, Sandrine Parageau, and John J. Thompson, 175–196. Leiden: Brill, 2016.

Martín Abad, Julián. "Crecimiento de la colección de Manuscritos de la Biblioteca Nacional en el siglo XIX: Breves apuntes para una historia necesaria." *Boletín de la ANABAD* 42, no. 1 (1992).

Martín-Valdepeña Yagüe, Elisa. "La reina María Luisa de Parma y la Junta de Damas de la Sociedad Económica Matritense." In *La época de Carlos IV (1788–1808): Actas del IV Congreso Internacional de la Sociedad Española de Estudios del Siglo XVIII*, edited by Elena de Lorenzo Álvarez, 749–62. Madrid: Ediciones Trea, 2009.

Martínez Martín, A. Jesús, ed. *Historia de la edición en España 1836–1936*. Madrid: Marcial Pons, 2001.

Mayordomo Pérez, Alejandro, and Luis Miguel Lázaro Lorente. *Escritos pedagógicos de la Ilustración*. Madrid: Centro de Publicaciones, MEC, 1989.

Mazzotti, Massimo. "Maria Gaetana Agnesi: Mathematics and the Making of the Catholic Enlightenment." *Isis* 92, no. 4 (December 2001): 657–83.

Mazzotti, Massimo. "Newton for Ladies: Gentility, Gender and Radical Culture." *British Journal for the History of Science* 37, no. 2 (June 2004): 119–46.

Mazzotti, Massimo. *The World of Maria Gaetana Agnesi, Mathematician of God*. Baltimore: Johns Hopkins University Press, 2007.

McCarthy, William. "Mother of All Discourses: Anna Barbault's Lessons for Children." In Ruwe, *Culturing the Child*, 85–111.

McClure, Ruth K. *Coram's Children: The London Foundling Hospital in the Eighteenth-Century*. New Haven: Yale University Press, 1981.

McKendrick, Neil, John Brewer, and J. H. Plumb. *The Birth of a Consumer Society: The Commercialization of Eighteenth-Century England*. Bloomington: Indiana University Press, 1982.

McMullen, Roy. "Introduction." In *Botany: A Study of Pure Curiosity; Botanical Letters and Notes towards a Dictionary of Botanical Terms by Jean-Jacques Rousseau*. London: Felix Gluck Press, 1979.

Medri, Sante, ed. *Giusseppe Compagnoni: Un intellettuale tra giacobinismo e restaurazione*. Bologna: Edizioni Analisi, 1993.

Messbarger, Rebecca. *The Century of Women: Representations of Women in Eighteenth-Century Italian Public Discourse*. Toronto: University of Toronto Press, 2002.

Messbarger, Rebecca. *The Lady Anatomist: The Life and Work of Anna Morandi Manzolini*. Chicago: Chicago University Press, 2010.

Messbarger, Rebecca Marie, and Paula Findlen, eds. *The Contest for Knowledge: Debates over Women's Learning in Eighteenth-Century Italy*. Chicago: University of Chicago Press, 2005.

Mestre, Antonio. *Apología y crítica de España en el siglo XVIII*. Madrid: Marcial Pons, 2003.

Milam, Erika Lorraine, and Robert A. Nye. "An Introduction to *Scientific Masculinities*." *Osiris* 30, no. 1 (January 2015): 1–14.

Miley, Frances, and Andrew Read. "Go Gentle Babe: Accounting and the London Foundling Hospital, 1757–97." *Accounting History* 21, nos. 2–3 (May 2016): 167–84.

Miranda, Pedro Álvarez de. "¿Una niña en la Academia? El caso de Rosario Cepeda y su orgullo padre." *Boletín de la Real Academia Española* 82 (2002): 39–45.

Mó Romero, Esperanza, and Margarita Eva Rodríguez García. "Educar: ¿A quién y para qué?" In Deusa, *Historia de las mujeres*, 729–56.

Morant, Isabel. "Reasons for Education: New Echoes of the Polemic." In Jaffe and Franklin Lewis, *Eve's Enlightenment*, 51–61.

Morant, Isabel, ed. *Historia de las mujeres en España y América Latina: El mundo moderno*. Vol. 2. Madrid: Cátedra, 2005.

Moreno de las Heras, M. "The Family of the Duques de Osuna." In Pérez Sánchez and Sayre *Goya and the Spirit of Enlightenment*, 38–40.

Mornet, Daniel. "Les enseignements des bibliothèques privées (1750–1780)." *Revue d'Histoire Littéraire de France* 18 (1910): 449–96.

Müller-Wille, Staffan. "Linnaeus and the Love Lives of Plants." In *Reproduction: From Antiquity to the Present Day*, edited by Nick Hopwood, R. Flemming, and L. Kassell, 305–18. Cambridge, UK: Cambridge University Press, 2018.

Müller-Wille, Staffan. "Reproducing Species." In *The Secrets of Generation: Reproduction in the Long Eighteenth Century*, edited by Raymond Stephanson and Darren N. Wagner, 37–58. Toronto: Toronto University Press, 2015.

Mullin, John. "Gendered Knowledge, Gendered Minds: Women and Newtonianism, 1690–1760)." In *A Question of Identity: Women, Science and Literature*, edited by Marina Benjamin, 41–58. New Brunswick, NJ: Rutgers University Press, 1993.

Myers, Greg. "Science for Women and Children: The Dialogue of Popular Science in the Nineteenth Century." In *Nature Transfigured: Science and Literature, 1700–1900*, edited by John Cristie and Sally Shuttleworth, 170–200. Manchester: Manchester University Press, 1989.

Myers, Mitzi. "Impeccable Governesses, Rational Dames and Moral Mothers: Mary Wollstonecraft and the Female Tradition in Georgian Children's Books." *Children's Literature* 14 (1986): 31–59.

Navarro, Victor, and William Eamon, eds. *Beyond the Black Legend: Spain and the Scientific Revolution*. Valencia: Universitat de València-CSIC, 2007.

Navascués, Pedro. "La Alameda de Osuna: Una villa suburbana." *Pro-Arte* 2 (1975): 7–26.

Negrín Fajardo, Olegario. *Ilustración y educación: La Sociedad Económica Matritense*. Madrid: Editora Nacional, 1984.

Nieto-Galán, Agustí. "The French Nomenclature in Spain: Critical Points, Rhetorical Arguments and Practical Uses." In Bensaude-Vincent and Abbri, *Lavoisier in European Context*, 173–91.

Nieto-Galán, Agustí. "The History of Science in Spain: A Critical Overview." *Nuncius* 23, no. 2 (January 2008): 211–36.

Nieto-Galán, Agustí. *Science in the Public Sphere: A History of Lay Knowledge and Expertise*. London: Routledge, 2016.

O'Brien, Karen. "Sexual Distinctions and Prescriptions." In Knott and Taylor, *Women, Gender and the Enlightenment*, 3–7.

O'Brien, Karen. *Women and Enlightenment in Eighteenth-Century Britain*. Cambridge, UK: Cambridge University Press, 2009.

O'Connor, Ralph. "Reflection on Popular Science in Britain: Gender, Categories and Historians." *Isis* 100, no. 2 (June 2009): 333–45.

O'Malley, Andrew. *The Making of the Modern Child: Children's Literature and Childhood in the Late Eighteenth Century*. New York: Routledge, 2003.
Orland, Barbara. "Enlightened Milk: Reshaping Bodily Substance into a Chemical Object." In Klein and Spary, *Materials and Expertise*, 163–97.
Ortega, Margarita. "El siglo XVIII: Introducción." In *Las mujeres de Madrid como agentes de cambio social*, edited by Margarita Ortega, 3–55. Madrid: Instituto Universitario de Estudios de la Mujer, 1995.
Ortega, Margarita. "La educación de la mujer en la ilustración española." *Revista de la Educación* (1988): 303–26.
Outram, Dorinda. *The Enlightenment*. Cambridge, UK: Cambridge University Press, 2013.
Palacios, Emilio. *La mujer y las letras en la España del siglo XVIII*. Madrid: Ediciones Laberinto, 2002.
Palma García, Dolores. "Las escuelas patrióticas creadas por la Sociedad Económica Matritense de Amigos del País en el siglo XVIII." *Cuadernos de historia moderna y contemporánea* 5 (1984): 37–56.
Papanelopoulou, Faidra, Agustí Nieto-Galán, and Enrique Perdriguero, eds. *Popularizing Science and Technology in the European Periphery, 1800–2000*. Aldershot: Ashgate, 2009.
Paquette, Gabriel B. "State-Civil Society Cooperation and Conflict in the Spanish Empire: The Intellectual and Political Activities of the Ultramarine Consulados and Economic Societies, c. 1780–1810." *Journal of Latin American Studies* 39, no. 2 (May 2007): 263–98.
Pardo-Tomás, José, and Álvar Martínez. "The Ignorance of Midwives: The Role of Cergymen in Spanish Enlightenment Debates on Birth Care." In *Medicine and Religion in Enlightenment Europe*, edited by Ole Peter Grell and Andrew Cunningham, 49–62. Aldershot: Ashgate, 2007.
Parra, Dolores, and Fernando Pelayo. "Christian Herrgen y la institucionalización de la mineralogía en Madrid." *Asclepio* 48 (1996): 163–81.
Pearson, Jacqueline. *Women's Reading in Britain, 1750–1835*. Cambridge, UK: Cambridge University Press, 1999.
Peiffer, Jeanne. "La litterature scientifique pour les femmes au xviii siècle." In *Sexe et genre: De la hierarchie entre les sexes*, edited by Marie-Claude Hurtig, Michele Kail, and Helene Rouch, 138–46. Paris: Centre National de la Recherche Scientifique, 1991.
Peña Soler, Álvaro Ruiz de la. "Ilustración, lectura y juventud en la España del XVIII." In Infantes, Lopez, and Botrel, *Historia de la edición y de la lectura*, 492–99.
Perdiguero-Gil, Enrique. "Popularizing Medicine during the Spanish Enlightenment." In *The Popularization of Medicine, 1650–1850*, edited by Roy Porter, 160–93. London: Routledge, 1992.
Pérez Cantó, Pilar. "Las españolas en la vida colonial." In Deusa, *Historia de las mujeres*, 2:525–54.
Pérez Cantó, Pilar, and Rocío de la Nogal. "Las mujeres en la arena pública." In Deusa, *Historia de las mujeres*, 757–89.
Pérez Carrión, Guillermo. "Casual Poverty in the Spanish Enlightenment: Josefa Amar y Borbón and the Real Sociedad Económica Aragonesa de Amigos del País." *Dieciocho: Hispanic Enlightenment* 26, no. 2 (2003): 265–93.
Pérez Moreda, Vicente. *La infancia abandonada en España (siglos XVI–XX)*. Madrid: Real Academia de la Historia, 2005.
Pérez Sánchez, Alfonso E., and Eleanor A. Sayre, eds. *Goya and the Spirit of Enlightenment*. Boston: Museum of Fine Arts, 1989.

Perkins, John. "Chemistry Courses, the Parisian Chemical World and the Chemical Revolution, 1770–1790." *Ambix* 57, no. 1 (March 2010): 27–47.

Peset, José Luis. "Enlightenment and Renovation in the Spanish University." In Feingold and Navarro-Brotons, *Universities and Science in the Early Modern Period*, 231–39.

Peset, José Luis. "La universidad de Alcalá, la mitra y la corona." *Miscelanea Alfonso IX* 43 (2003): 33–43.

Phillips, Denise. "Experimentation in the Agricultural Enlightenment: Place, Profit and Norms of Knowledge-Making in Eighteenth-Century Germany." In "Expectations and Utility in Eighteenth-Century Knowledge Economies," edited by Larry Stewart and Kelly Whitmer. Special issue, *Notes and Records: The Royal Society Journal of the History of Science* 72, no. 2 (June 2018): 159–72.

Pimentel, Juan, and José Pardo-Tomás. "And Yet, We Were Modern: The Paradoxes of Iberian Science after the Grand Narratives." *History of Science* 55, no. 2 (June 2017): 133–47.

Pinault-Sørensen, Madeleine. "Les planches du Spectacle de la nature de l'abbé Pluche." In Gevrey, Boch, and Haquette, *Écrire la nature au XVIIIe siècle*, 441–59.

Porter, Roy. "The Environment and the Enlightement: The English Experience." In Daston and Pomata, *Faces of Nature*, 17–38.

Porter, Roy S., and Mikuláš Teich, eds. *The Enlightenment in National Context*. Cambridge, UK: Cambridge University Press, 1981.

Porter, Theodore M. "Making Things Quantitative." *Science in Context* 7, no. 3 (Autumn 1994): 389–407.

Porter, Theodore M. *Trust in Numbers: The Pursuit of Objectivity in Science and Public Life*. Princeton, NJ: Princeton University Press, 1995.

Pradells, Jesús. "Juan Bautista Virio (1753–1837): Experiencia europea y reformismo económico en la España ilustrada." *Revista de historia moderna* 8–9 (1988–1989): 233–71.

Prunean, Alexandra. "Commercial Strategies in Paratextual Features of Late Eighteenth-Century Children's Books." PhD diss., Universidad Autónoma de Barcelona, 2016.

Rankin, Alisha. *Panaceas' Daughters: Noblewomen as Healers in Early Modern Germany*. Chicago: University of Chicago Press, 2013.

Rauch, Alan. "Mentoria: Women, Children and the Structures of Science." In *Nineteenth-Century Contexts* 27, no. 4 (December 2005): 335–51.

Redlich, Fritz. "Science and Charity: Count Rumford and His Followers." *International Review of Social History* 16, no. 2 (August 1971): 184–216.

Remón Fernández, Juan F. "The Alameda of the Duchess of Osuna: A Garden of Ideas." *Journal of Garden History* 13, no. 4 (1993): 224–40.

Rendall, Jane. "Gender, Race and the Progress of Civilization: Introduction." In Knott and Taylor, *Women, Gender and Enlightenment*, 70–74.

Reuter, A., "Family of the Duke and Duchess of Osuna, 1787–1788." In Tomlinson, *Goya: Images of Women*, 162–64.

Reynolds, Kimberley. "Changing Families in Children's Fiction." In Grenby and Immel, *Cambridge Companion*, 193–208.

Richardson, Tim. *The Arcadian Friends: Inventing the English Landscape Garden*. London: Bantam, 2007.

Rivest, Justin. "Testing Drugs and Attesting Cures: Pharmaceutical Monopolies and Military Contracts on Eighteenth-Century France." In "Testing Drugs and Trying Cures." Special issue, *Bulletin of the History of Medicine* 91, no. 2 (Summer 2017): 362–90.

Roberts, Lissa. "Exploring Global History through the Lens of History of Chemistry: Materials, Identities and Governance." *History of Science* 54, no. 4 (December 2016): 335–61.

Roberts, Lissa. "Going Dutch: Situating Science in the Dutch Enlightenment." In Clark, Golinski, and Schaffer, *Sciences in Enlightened Europe*, 350–88.

Roberts, Lissa. "Instruments of Science and Citizenship: Science Education for Dutch Orphans during the Late 18th Century." *Science and Education* 21, no. 2 (February 2012): 157–77.

Roberts, Lissa. "Introduction." In *The Mindful Hand: Inquiry and Invention from the Late Renaissance to Early Industrialization*, edited by Lissa Roberts, Simon Schaffer, and Peter Dear, 1–10. Amsterdam: KNAW, 2007.

Roberts, Lissa. "Practicing *Oeconomy* during the Second Half of the Eighteenth Century: An Introduction." *History and Technology* 30, no. 3 (2014): 133–48.

Roberts, Lissa. "Situating Science in Global History: Local Exchanges and Networks of Circulation." *Itinerario* 33, no. 1 (March 2009): 9–30.

Roberts, Lissa, and Simon Werrett. "Introduction: A More Intimate Acquaintance." In Roberts and Werrett, *Compound Histories*, 1–32.

Roberts, Lissa, and Simon Werrett, eds. *Compound Histories: Materials, Governance and Production, 1760–1840*. Leiden: Brill, 2017.

Rodríguez-Moñino, Antonio. *La Imprenta de Don Antonio de Sancha (1771–1790)*. Madrid: Castalia, 1971.

Rothschild, Emma. *Economic Sentiments: Adam Smith, Condorcet, and the Enlightenment*. Cambridge, MA: Harvard University Press, 2001.

Ruiz Torres, Pedro, "Reformismo e ilustración." In *Historia de España*, edited by Josep Fontana and Ramón Villares, 5:610–15. Barcelona: Marcial Pons, 2007.

Rusnock, Andrea. "Quantifying Infant Mortality in England and France, 1750–1800." In Jorland, Opinel, and Weisz, *Body Counts*, 65–86.

Russell, Terence M., and Anne Marie Thornton. *Gardens and Landscapes in the Encyclopédie of Diderot and D'Alembert: The Letterpress Articles and Selected Engravings*. Aldershot: Ashgate, 1999.

Ruwe, Donelle, ed. *Culturing the Child (1690–1914): Essays in Memory of Mitzi Myers*. Lanham, MD: Scarecrow Press, 2005.

Sánchez-Blanco, Francisco. *El absolutismo y las luces*. Madrid: Marcial Pons, 2002.

Sánchez-Blanco, Francisco. *La Ilustración en España*. Madrid: Akal Ediciones, 1997.

Sánchez García, Raquel. "La edición de libros infantiles y juveniles." In Martínez Martín, *Historia de la edición en España*, 337–54.

Sarasúa, Carmen. *Criados, nodrizas y amos: El servicio doméstico en la formación del mercado de trabajo madrileño, 1758–1868*. Madrid: Siglo Veintiuno de España Editores, 1994.

Sarasúa, Carmen. "The Hardest, Most Unpleasant Profession: The Work of Laundresses in Eighteenth-, Nineteenth- and Twentieth-Century Spain." In *A Social History of Spanish Labor: New Perspectives on Class, Politics and Gender*, edited by José Piqueras and Vicent Sanz-Rozalén, 64–90. New York: Berghahn Books, 2008.

Sarasúa, Carmen. "Technical Innovations at the Service of Cheaper Labour in Pre-Industrial Europe: The Enlightened Agenda to Transform the Gender Division of Labor in Silk Manufacturing." *History and Technology* 24, no. 1 (2008): 23–29.

Sarasúa, Carmen. "Una política de empleo antes de la industrialización: Paro, estructura de la ocupación y salarios en la obra de Campomanes." In *Campomanes y su obra económica*, edited by Francisco Comin and Pablo Martin, 171–91. Madrid: Instituto de Estudios Fiscales, 2005.

Sarrailh, Jean. "D. Diego Clemencín." *Bulletin Hispanique* 24 (1922): 125–30.

Sarrailh, Jean. *La España Ilustrada de la segunda mitad del siglo XVIII*. Mexico City: Fondo de Cultura Económica, 1957.

Savini, Marcello. *Un abate libertino, Le memoire autobiografiche e altri scritti di Giussepe Compagnoni*. Lugo: Banca di Romagna, 1988.

Schabas, Margaret, and Neil Di Marchi. "Introduction to Oeconomies in the Age of Newton." In "Oeconomies in the Age of Newton," edited by Schabas and Di Marchi. Special issue, *History of Political Economy* 35 (2003): 1–13.

Schaffer, Simon. "Enlightenment Brought Down to Earth." *History of Science* 41, no. 3 (September 2003): 257–68.

Schaffer, Simon, Lissa Roberts, Kapil Raj, and James Delbourgo, eds. *The Brokered World: Go-Betweens and Global Intelligence, 1770–1820*. Sagamore Beach: Science History, 2009.

Schiebinger, Londa. *The Mind Has No Sex? Women in the Origins of Modern Science*. Cambridge, MA: Harvard University Press, 1989.

Schiebinger, Londa. "Women of Natural Knowledge." In Daston and Park, *Early Modern Science*, 192–205.

Schiebinger, Londa, and Claudia Swan. *Colonial Botany: Science, Commerce and Politics in the Early Modern World*. Philadelphia: University of Pennsylvania Press, 2005.

Sebastiani, Silvia. "'Race,' Women and Progress in the Scottish Enlightenment." In Knott and Taylor, *Women, Gender and Enlightenment*, 75–96.

Secord, James. "Knowledge in Transit." *Isis* 95, no. 4 (December 2004): 654–72.

Secord, James. "Newton in the Nursery: Tom Telescope and the Philosophy of Tops and Balls (1761–1838)." *History of Science* 23, no. 2 (June 1985): 127–51.

Serrano, Elena. "Making *Oeconomic* People: The Spanish Magazine of Agriculture and Arts for Parish Rectors (1797–1808)." *History and Technology* 30, no. 3 (2014): 149–76.

Serrano, Elena. "Sex and Prisons: Women and Spanish Penitentiary Reform, 1787–1808." In "The Sexes and the Sciences," edited by D. N. Wagner and J. Wharton. Special issue, *Journal for Eighteenth-Century Studies* 42, no. 4 (December 2019): 501–17.

Serrano, Elena. "The *Spectacle de la Nature* in Eighteenth-Century Spain: From French Households to Spanish Workshops." *Annals of Science* 69 (2012): 257–82.

Serrano, Elena. "Spreading the Revolution: Guyton's Fumigating Machine in Spain. Politics, Technology, and Material Culture (1796–1808)." In Roberts and Werrett, *Compound Histories*, 106–30.

Serrano, Rafael. "Técnicas agrícolas y zootecnia." In García Ballester, *Historia de la ciencia*, 4:603–30.

Shapin, Steven. "The Image of the Man of Science." In *The Cambridge History of Science*, vol. 4, *Eighteenth-Century Science*, edited by Roy Porter, 159–83. Cambridge, UK: Cambridge University Press, 2003.

Shefrin, Jill. "'Governesses to Their Children': Royal and Aristocratic Mothers Educating Daughters in the Reign of George III." In Immel and Witmore, *Childhood and Children's Books*, 181–211.

Shefrin, Jill. "Make It a Pleasure and Not a Task: Educational Games for Children in Georgian England." *Princeton University Library Chronicle* 40, no. 2 (Winter 1999): 251–75.

Sherwood, Joan. *Infection of the Innocents: Wet Nurses, Infants and Syphilis in France*. Montreal: McGill-Queen's University Press, 2010.

Sherwood, Joan. *Poverty in Eighteenth-Century Spain: The Women and Children of the Inclusa.* Toronto: University of Toronto Press, 1988.

Shteir, Ann B. *Cultivating Women, Cultivating Science: Flora's Daughters and Botany in England, 1760 to 1860.* Baltimore: Johns Hopkins University Press, 1996.

Shteir, Ann B. "Priscilla Wakefield's Natural History Books." In *From Linnaeus to Darwin: Commentaries on the History of Biology and Geology*, edited by Alwayne Wheeler and James H. Price, 29–36. London: Society for the History of Natural History, 1985.

Sluga, Glenda, and Carolyn James, eds. *Women, Diplomacy and International Politics since 1500.* New York: Routledge, 2016.

Shuttleworth, Sally. "Constructing Scientific Communities: Citizen Science." *Lancet* 385, no. 9978 (2015): 2568.Smith, Theresa Ann. *The Emerging Female Citizen: Gender and Enlightenment in Spain.* Berkeley: University of California Press, 2006.

Smith, Theresa Ann. "Writing Out of the Margins: Women, Translation, and the Spanish Enlightenment." *Journal of Women's History* 15, no. 1 (Spring 2003): 116–43.

Soubeyroux, Jacques. "Niveles de alfabetización en la España del siglo XVIII: Primeros resultados de una encuesta en curso." *Revista de Historia Moderna: Anales de la Universidad de Alicante* 5 (1985): 159–72.

Sparey, Victoria. "Identity-Formation and the Breastfeeding Mother in Renaissance Generative Discourses and Shakespeare's Coriolanus." *Social History of Medicine* 25, no. 4 (October 2012): 777–94.

Spary, Emma C. *Feeding France: New Sciences of Food, 1760–1815.* Cambridge, UK: Cambridge University Press, 2014.

Spary, Emma C. "Liqueurs and the Luxury Marketplace in Eighteenth-Century Paris." In Klein and Spary, *Materials and Expertise*, 225–57.

Sperling, Jutta Gisela. *Medieval and Renaissance Lactations: Images, Rhetoric, Practices.* Burlington, VT: Ashgate, 2013.

Stapelbroek, Koen, and Jani Marjanen, eds. *The Rise of Economic Societies in the Eighteenth Century: Patriotic Reform in Europe and North America.* Basingstoke: Palgrave Macmillan, 2012.

Stewart, Larry, and Kelly J. Whitmer. "Expectations and Utility in Eighteenth-Century Knowledge Economies." In "Expectations and Utility in Eigtheenth-Century Knowledge Economies," edited by Stewart and Whitmer. Special issue, *Notes and Records: The Royal Society Journal of the History of Science* 72, no. 2 (June 2018): 111–17.

Sullivan, Constance. "Constructing Her Own Tradition: Ideological Selectivity in Josefa Amar y Borbón's Representation of Female Models." In *Recovering Spain's Feminist Tradition*, edited by Lisa Vollendorf, 142–59. New York: Modern Languages Association of America, 2001.

Sullivan, Constance. "Josefa Amar y Borbon and the Royal Aragonese Economic Society (with documents)." *Dieciocho: Hispanic Enlightenment* 15 (1992): 1–2 and 95–148.

Sussman, George D. *Selling Mother's Milk: The Wet-Nursing Business in France 1715–1914.* Urbana: University of Illinois Press, 1982.

Sutton, Geoffrey V. *Science for a Polite Society: Gender, Culture, and the Demonstration of Enlightenment.* Oxford: Westview, 1995.

Tarbuck, Derya Gurses. "Exercises in Women's Intellectual Sociability in the Eighteenth Century: The Fair Intellectual Club." *History of European Ideas* 41, no. 3 (2015): 375–86.

Terrall, Mary. "Gender and Early Modern Science Cluster. Introduction." In "Gender and

Early Modern Science Cluster," edited by Mary Terrall. Special issue, *Configurations* 3, no. 2 (Spring 1995): 135–38.

Terrall, Mary. "Gendered Spaces, Gendered Audiences: Inside and Outside the Paris Academy of Sciences." *Configurations* 3, no. 2 (Spring 1995): 207–32.

Terrall, Mary. "Masculine Knowledge, the Public Good, and the Scientific Household of Réaumur." *Osiris* 30, no. 1 (January 2015): 182–202.

Terrall, Mary. "Metaphysics, Mathematics, and the Gendering of Science in Eighteenth-Century France." In Clark, Golinski, and Schaffer, *Sciences in Enlightened Europe*, 246–71.

Terrall, Mary. "Natural Philosophy for Fashionable Readers." In *Books and the Sciences in History*, edited by Marina Frasca-Spada and Nick Jardine, 239–53. Cambridge, UK: Cambridge University Press, 2000.

Terreros, Esteban de. *Diccionario castellano con las voces de ciencias y artes y sus correspondientes en las tres lenguas francesa, latina e italiana*. 4 vols. Madrid: Viuda de Ibarra, 1786–1793.

Thompson, Carl. "Women Travelers, Romantic-Era Science and the Banksian Empire." *Notes and Records: The Royal Society Journal of the History of Science* 73, no. 4 (December 2019): 431–55.

Toma, Stefania de. "Giuseppe Compagnoni e la divulgazione scientifica nella Chimica per le donne." In *La letteratura italiana a Congresso: Bilanci e prospettive del decennale (1996–2006), Atti del X Congresso dell'ADI*, edited by R. Cavalluzzi et al., 619–28. Lecce: Pensa multimedia, 2008.

Tomaselli, Sylvana. "Civilization, Patriotism and Enlightened Histories of Woman." In Knott and Taylor, *Women, Gender and Enlightenment*, 117–35.

Tomaselli, Sylvana. "Moral Philosophy and Population Questions in Eighteenth-Century Europe." *Population and Development Review* 14, supplement (1988): 7–29.

Tomlison, Janis, ed. *Goya: Images of Women*. Washington, DC: National Gallery of Art, 2002.

Topham, Jonathan R. "Historizing Popular Science." *Isis* 100, no. 2 (June 2009): 310–18.

Topham, Jonathan R. "Publishing Popular Science." In Fyfe and Lightman, *Science in the Marketplace*, 135–68.

Topham, Jonathan R., "Rethinking the History of Science: Popularization/Popular Science." In Papanelopoulou, Nieto-Galán, and Perdiguero, *Popularizing Science and Technology*, 1–20.

Topham, Jonathan R. "The Scientific, the Literary and the Popular: Commerce and the Reimagining of the Scientific Journal in Britain, 1813–1825." *Notes and Records of the Royal Society of London* 70, no. 4 (December 2016): 305–24.

Trinkle, Dennis. "Nöel-Antoine Pluche's *Le spectacle de la nature:* An Encyclopaedic Best Seller." *Studies on Voltaire and the Eighteenth Century* 358 (1997): 93–134.

Tröhler, Ulrich. "Quantifying Experience and Beating Biases: A New Culture in Eighteenth-Century British Clinical Medicine." In Jorland, Opinel, and Weisz, *Body Counts*, 19–50.

Ugarte Blanco, Juana. "Economía en Espagne au 18e siècle: Fonctionnements discursifs et sociaux." *Mots* 15 (1987): 93–109.

Urzainqui, Inmaculada. *"Catalín." De Rita Barrenecha y otras voces de mujeres en el siglo XVIII*. Vitoria-Gasteiz: Santamaría, 2006.

Urzainqui, Inmaculada. "Nuevas propuestas a un público femenino." In Infantes, López, and Botrel, *Historia de la edición y la lectura*, 481–98.

Urzainqui, Inmaculada. "Un nuevo instrumento cultural: la prensa periódica." In Álvarez Barrientos, López, and Urzainqui, *La República de las letras*, 125–216.

Valverde, Nuria. *Actos de precision: instrumentos científicos, opinión pública y economía moral en la Ilustración española*. Madrid: CSIC, 2007.

Varey, John E. *Cartelera de los títeres y otras diversiones populares de Madrid: 1758–1840. Estudio y documentos*. Madrid: Támesis, 1995.

Vázquez Madruga, María Jesús. *María Isidra Quintina de Guzmán y la Cerda, la doctora de Alcalá*. Madrid: Centro Asesor de la Mujer, 1999.

Vetter, Jeremy. "Introduction: Lay Participation in the History of Scientific Observation." *Science in Context* 24, no. 2 (June 2011): 127–41.

Vicente, Marta V. *Clothing the Spanish Empire: Families and the Calico Trade in the Early Modern Atlantic World*. Basingstoke: Palgrave Macmillan, 2006.

Vicente, Marta V. "Textual Uncertainties: The Written Legacy of Women Entrepreneurs in Eighteenth-Century Barcelona." In *Women, Texts and Authority in the Early Modern Spanish World*, edited by Marta V. Vicente and Luis R. Corteguera, 182–92. Aldershot: Ashgate, 2003.

Vickery, Amanda. *Behind Closed Doors: At Home in Georgian England*. New Haven: Yale University Press, 2009.

Viñao Frago, Antonio. "El libro escolar." In Martínez Martín, *Historia de la edición en España*, 309–36.

Vismann, Cornelia. *Files: Law and Media Technology*. Abridged and translated by Geoffrey Winthrop-Young. Stanford, CA: Stanford University Press, 2008.

Wharton, Joanna. *Material Enlightenment: Women Writers and the Science of Mind, 1770–1830*. Vol. 1. Suffolk: Boydell & Brewer, 2018.

Werrett, Simon. "Household Economy and Chemical Inquiry." In Roberts and Werrett, *Compound Histories*, 35–56.

Werrett, Simon. "The Sociomateriality of Waste and Scrap Paper in Eighteenth-Century England." In Bittel, Leong, and von Oertzen, *Working with Paper*, 46–59.

Werrett, Simon. *Thrifty Science: Making the Most of Materials in the History of Experiments*. Chicago: Chicago University Press, 2019.

Werrett, Simon. "Watching the Fireworks: Early Modern Observation of Natural and Artificial Spectacles." *Science in Context* 24, no. 2 (June 2011): 167–82.

Wickes, Ian G. "A History of Infant Feeding. Part III: Eighteenth and Nineteenth Century Writers." *Archives of Disease in Childhood* 139 (June 1953): 332–40.

Williamson, Tom. *Polite Landscapes: Gardens and Society in Eighteenth-Century England*. Baltimore: Johns Hopkins University Press, 1995.

Zavala, Iris M., ed. *Breve historia feminista de la literatura española*. Barcelona: Anthropos Editorial, 1997.

Zorrozua, Maria del Pilar. *Escritoras de la Ilustración española (1759–1808)*. Bilbao: Dpto. Publicaciones Universidad Deusto, 1999.

INDEX

Note: Page numbers in italics indicate figures.